Desalination - Challenges and Opportunities

Edited by
Mohammad Hossein Davood Abadi Farahani,
Vahid Vatanpour and Amir Hooshang Taheri

Published in London, United Kingdom

IntechOpen

Supporting open minds since 2005

Desalination - Challenges and Opportunities
http://dx.doi.org/10.5772/intechopen.77449
Edited by Mohammad Hossein Davood Abadi Farahani, Vahid Vatanpour and Amir Hooshang Taheri

Contributors
Mona Abdel-Fatah, Ghada Al-Basedi, S.M. Fakhrul Islam, Zahurul Karim, Lucy Lunevich, Vel Murugan, Qian Chen, Muhammad Burhan, Muhammad Wakil Shahzard, Raid Alrowais, Doskhan Ybyraiymkul, Faheem Hassan Akhtar, Yong Li, Kim Choon Ng, Mohammad Hossein Davood Abadi Farahani, Reza Katal, Teo Ying Shen, Iman Jafari, Saeid Masudy-Panah

First published in London, United Kingdom, 2020 by IntechOpen
IntechOpen is the global imprint of INTECHOPEN LIMITED, registered in England and Wales, registration number: 11086078, 7th floor, 10 Lower Thames Street, London, EC3R 6AF, United Kingdom
Printed in Croatia

British Library Cataloguing-in-Publication Data
A catalogue record for this book is available from the British Library

Additional hard and PDF copies can be obtained from orders@intechopen.com

Desalination - Challenges and Opportunities
Edited by Mohammad Hossein Davood Abadi Farahani, Vahid Vatanpour and Amir Hooshang Taheri
p. cm.
Print ISBN 978-1-78984-738-3
Online ISBN 978-1-78984-739-0
eBook (PDF) ISBN 978-1-83962-703-3

For EU product safety concerns:
IN TECH d.o.o., Prolaz Marije Krucifikse Kozulić 3, 51000 Rijeka, Croatia, info@intechopen.com or visit our website at intechopen.com.

Meet the editors

Dr Mohammad H.D.A. Farahani obtained his BSc and MSc in chemical engineering in 2012 and 2014, respectively. He received his Ph.D. degree in chemical and biomolecular engineering (membrane technology) in 2018 from the National University of Singapore. His Ph.D. thesis received "The Outstanding Postgraduate Research Thesis Award" from the American Institute of Chemical Engineers in 2018. He also received the prestigious title of "Innovator Under 35 by MIT Technology Review" in 2020. His research over the past 9 years has resulted in 2 patents, more than 20 peer-reviewed articles that have been cited almost 700 times, a book chapter, and many international presentations and talks as invited and keynote speaker. Dr Farahani is currently the co-founder and CEO of SEPPURE. SEPPURE creates industrial-scale solvent-resistant nanofiltration membranes to sustainably separate chemical mixtures at a molecular level with minimal energy use.

Dr Vahid Vatanpour was born in Sarein (Iran) in 1983. He obtained his BSc and MSc degrees in applied chemistry from the University of Tabriz and his PhD in 2012 from the Department of Chemical Engineering, Razi University, Iran. He was assistant professor of applied chemistry in the Department of Chemistry of the Kharazmi University. Since 2016 he has been an associate professor. His research interests are focused on membrane science and technology for water and wastewater treatment, especially fabrication of antifouling polymeric membranes. He has published over 130 papers in refereed journals.

Dr Taheri started to work on membrane in 2005 when he joined the Razi University to receive his MSc in chemical engineering. He obtained his Ph.D. at Nanyang Technological University, working in the highly respected Singapore Membrane Technology Centre (SMTC) under the supervision of Professor Anthony G. Fane. He has two patent applications, more than 25 publications/presentations with +750 citations in the field of membrane science and technology. His current interests are membranes applied to vegetable oil, the pharmaceutical and petrochemical industries, with a focus on the sustainability aspects of membrane technology as a CTO and co-founder of SEPPURE.

Contents

Preface

Today, freshwater supply has become a significant concern worldwide as a result of population growth and a high water demand. According to new estimates, there are more than one billion people without access to safe drinking water, and about 2.3 billion people (41% of the world's population) live in water-deprived areas. In many cases, solutions such as water transportation or dam construction are not sufficient and useful ways to cope with the increasing demand for water and limited resources. Traditional freshwater resources such as lakes, rivers, or groundwater are underutilized or inappropriately harvested, and as a result, these resources are either depleted or saline. Water and energy are necessary for life on earth and to sustain the modern world. In many parts of the developed world, the control and exploitation of water and energy has driven economic development and progress. In the developing world, many regions suffer from shortages of freshwater and energy supplies. The United Nations Environment Program (UNEP) stated that one third of the world's population live in countries with insufficient freshwater to support the population. Consequently by 2025, two thirds of the world population will face water scarcity.

Drinking water of acceptable quality has become a scarce commodity. The World Health Organization estimates that over a billion people lack access to purified drinking water and the vast majority of these people are living in rural areas where the low population density and remote locations make it difficult to install traditional clean water solutions. Unfortunately, in addition to being scarce, freshwater resources are also unevenly distributed worldwide. As countries develop and cities expand, new and limited water resources are introduced for daily water supply. An increase in global warming, a reduction in the amount of precipitation, a significant reduction in groundwater levels, an increase in population, the emergence of new cities and metropolises, and the spread of water pollution, have raised the alarm for the future of water on earth and made it one of the most important challenges for many countries these days.

Hence, water shortage is becoming a major concern all around the world due to limited freshwater resources as well as the high cost of freshwater transportation from freshwater-rich areas to arid areas. As a result, solutions such as water recycling and desalination of saline or brackish water are being introduced and emerging worldwide as alternative ways of supplying water.

Desalination of seawater is known to be one of mankind's earliest forms of water treatment, and it has become one of the most sustainable alternative solutions to provide freshwater for many communities and industrial sectors. This plays a crucial role in socioeconomic development in a number of developing countries, especially in water-stressed regions such as Africa, Pacific Asia, and countries in the Middle East. Hence, the increase in population together with the industrial and agricultural development in emerging countries will rapidly accelerate the deterioration and depletion of the available freshwater resources. For the past few decades, the salinity of these water resources has been a major obstacle to their use for drinking or for use in industry and agriculture.

In terms of salinity, TDS (total dissolved solids), which is the total ionic concentration of dissolved minerals in the water, is defined to estimate the level of salts in water. It is proven that a high TDS concentration in water not only can affect human health but also could potentially cause other problems such as corrosion or scaling in pipes. Hence, the goal of desalination is to change high salinity water such as seawater or brackish water into freshwater for drinking, agriculture, or industrial uses.

Despite the various benefits of water desalination, these processes are accompanied by severe challenges, including adverse environmental effects. For example, seawater desalination can cause serious damage to the environment if the plants don't comply with the requirements for their water intake and also their brine disposal. The high volume and high salinity of the disposed brines can significantly change the marine ecosystem in the area and destroy marine life. Large commercial desalination plants that use fossil fuels are in use in most of the countries suffering from water shortages. For instance, a number of oil-rich countries use fossil fuels to supplement the energy for water desalination supply. In contrast people in many other areas of the world have neither the financial nor oil resources to allow them to develop in a similar manner. The production of 1000 m3 per day of freshwater requires 10,000 tonnes of oil per year, which can be considered a highly significant energy consumption, as it involves a current energy expense that few of the water-short areas of the world can afford. Recently, the utilization of renewable sources (e.g., solar, biomass, wind, and geothermal) to drive desalination plants has emerged as a promising sustainable solution for freshwater supply in regions lacking energy supply. This may be especially significant in regions where water is needed and renewable resources are available such as Africa and the Middle East region. The conversion of solar radiation into direct utilization has been investigated for many years. Recently, attention has been directed towards improving the conversion efficiency of solar energy systems, desalination technologies, and their optimal coupling to make them economically viable for small and medium scale applications.

This book presents a summary for readers interested in the area of desalination, and its challenges and opportunities. It provides detailed information on the treatment and management of the desalination brine solution, the effect of climate change on food and water demand, and the planning and engineering of brackish and seawater desalination projects.

We are thankful to all researchers and engineers who shared their knowledge and expertise to gather information for this book.

Dr Mohammad Hossein Davood Abadi Farahani and Dr Amir Hooshang Taheri
Research and Development Department,
Seppure, Singapore

Dr Vahid Vatanpour
Faculty of Chemistry,
Kharazmi University,
Tehran, Iran

Chapter 1

An Overview on the Treatment and Management of the Desalination Brine Solution

Reza Katal, Teo Ying Shen, Iman Jafari, Saeid Masudy-Panah and Mohammad Hossein Davood Abadi Farahani

Abstract

Due to the increasing limitations of water resources, application of desalination plants is expanding. One of the constraints associated with desalination plant operation is the production of concentrated solution, which is known as brine and can lead to critical challenges in the environment due to its high level of salinity. In this regard, many different disposal options used recently to control and prevent the environmental issues may be caused by the brine. Evaporation ponds, surface water discharge, and deep well injection are considered as the most well-known options to properly dispose concentrated brine. However, the application of these methods is highly restricted by capital cost and their limited uses. The treatment methods vary in terms of their ability in organics removal and can be divided into three different conventional groups as biological, physicochemical, and oxidation. In recent years, more attention has been paid to membrane-based technologies due to their economic performance in recovering precious resources and providing potable water with high recovery rates. This book chapter provides some critical reviews on recent technologies including treatment operations and disposal options to manage concentrated solutions from desalination plants. Finally, electrodialysis, forward osmosis, and membrane distillation as emerging membrane processes are examined in this chapter.

Keywords: brine, management, treatment, disposal, metal recovery

1. Characteristics of brine solution

The by-product of seawater desalination is known as brine. Brine is extremely concentrated seawater that causes detrimental environmental impacts due to its high salinity and presence of multifarious contaminants. They include heavy metals, nutrients containing nitrogen and phosphorus [ammonia (NH_3), nitrate (NO_3^-), and phosphorus (K)] derivatives, organic contaminants in minute concentrations (hormone and endocrinal disruptors, pharmaceutical and personal care products, soluble microbial products, and incompletely degraded organics found

EC (mS/cm)	TDS (mg/L)	Ca^{2+} (mg/L)	Mg^{2+} (mg/L)	Na^{+} (mg/L)	K^{+} (mg/L)	Cl^{-} (mg/L)	SO_4^{2-} (mg/L)	HCO_3^{-} (mg/L)	PO_4^{3-} (mg/L)	Ref.
15.5	10927.7	959.4	378.5	2024	70.4	4817	2560.3	—	—	[16]
38.7	34,885	1855	1556	7359	241	14,428	8366	863	0.6	[17]
85.2	79,660	960	2867	25,237.28	781.82	41,890	6050	1829	—	[18]
76.8	57,400	521	1738	18,434	491	32,127	4025	—	2.5	[19]
—	80028.4	891.2	2877.7	24649.2	888	43661.5	6745.1	315.3	—	[20]

Table 1.
Constituents of brine from various desalination plants [15].

in wastewater effluents), and pathogenic microorganisms. The identification of such trace organic contaminants is cardinal because it only serves to indicate the increased propensity of incomplete removal in most wastewater treatment systems across the globe [1]. Such prolonged exposure to such organic contaminants can lead to incomprehensible and detrimental impacts on our ecology albeit present in trace amounts. Environments in close proximity to brine discharge feedwater could be ravaged ecologically and physicochemically because the concentration of such contaminants is multiplied several-fold [2].

The matrix of brine depends on a confluence of factors—quality of water sources, desalination processes the brine is subjected to, the permeate water grade, pretreatment unit processes, and chemicals employed during the cleaning-in-place [3]. Research has shown that acids, antiscalants, and biocides have a direct consequential effect on the equilibria of the dissolved constituents [4]. As a result, the matrix of brine can differ because of the constituents' concentration and characteristics due to the employment of chemicals during pre-RO treatment. Ersever et al. [5] found that the RO concentrate from a Californian water reclamation plant in California had trace copper (Cu), manganese (Mn), mercury (Hg), and selenium (Se) levels. Alkalinity presence was also identified in brine, ranging approximately from 500 to 1500 mg L^{-1} as $CaCO_3$, alongside with NH_3 at around 60–100 mg N L^{-1}, chloride amounting to 800–1000 mg L^{-1}, and sulfate at 1000–1500 mg L^{-1}. Yoon et al. [6], on the other hand, found that the concentrate from the RO treatment of farm animals' wastewater treatment contained NH_3, humic substances, NO_3^-, phosphate (PO_4^{3-}), and potassium (K). PO_4^{3-} concentration reached up to 40 mg L^{-1} when the feed concentrations were as low as 5 mg L^{-1} [7]. Gomes et al. found that the concentrates from the RO treatment from textile plants had high levels of chemical oxygen demand (COD) of up to 15,000 mgL^{-1} [8]. Subramani et al. also noted that the RO brine obtained from treating water produced during oil and gas production contained high concentrations of silica and total organic carbon superseding 250 and 60 mg L^{-1}, respectively [9].

The characteristics of RO brine from industrial areas, however, differ from municipalities. For example, groundwater treatment sites contaminated by mining activities contained high calcium, silica, and sulfate concentrations greater than 1000, 200, and 4500 mg L^{-1}, respectively [10]. Randall et al. noted that the brine conductivity levels from mining industries were hovering at 22,000, almost on par to electrical conductivity (EC) levels of desalination plants' RO concentrate [11]. Umar et al. posited that EC levels were almost at 25 mS cm^{-1} for brine solution from a municipal wastewater [12]. The high total dissolved solid (TDS) content is attributed to such high EC levels in brine [13]. Generation of brine from desalination of brackish groundwater was found to contain barium, calcium, silica, and sulfate [10, 14]. High concentration of these TDS causes scaling because the concentrations of barium sulfate ($BaSO_4$), calcium carbonate ($CaCO_3$), and calcium sulfate ($CaSO_4$) have saturated and exceeded their solubility limit (K_{sp}). This diminishes the permeate recovery of the RO process. The constituents of brine are shown in **Table 1**, which depicts the changes of the water quality. These variations occur because sources of influents are different, alongside with different design and operational parameters employed in the treatment processes.

2. Environmental impacts

Brine is defined as the waste by-product of desalination. A plethora of research studies have conducted environmental impact assessment studies of brine disposal on identified areas of concerns—marine, groundwater, and soil environments [21, 22]. The environmental issues are as multifaceted as they are multifarious. Experts associate brine disposal with a correlated increase in salt concentrations of water bodies which are recipients of disposed brine. Brines with high TDS detrimentally affect the marine benthic communities living in close proximity to brine discharge. Brine disposal is also esthetically unpleasing. Moreover, the presence of chemicals used for pretreatment and membrane cleaning and corrosive metals are of pivotal concern. Salinity, temperature, and chemical composition of brine are reasons why brine poses a threat to our environments. The salinity and temperatures of brine depend entirely on the unit treatment processes employed. Brine salinity hovers at about 55–70 g/L, about 1.5–2 times much higher than seawater. Further, the brine produced by thermal-based technologies fluctuates at about 30–40°C, 1.37–1.82 times higher than seawater temperature at 22° [23, 24].

The marine environments would not be harmed if brine was disposed from a sole desalination plant, but the collective discharge of brines from multiple plants operating close to each other for extended periods will. Numerous studies point to brine disposal as the core reason for osmotic balance disruption as it increases the salinity of marine life habitats. Living organisms will be depraved of water at a cellular level, and the increased salinity causes a turgor pressure decline [25, 26]. This may lead to the plausible eradication of species [27, 28]. Jenkins et al. propounded that several marine species could be detrimentally affected by salinity permutations of only 2–3 g/L, though some species may be sturdier to such salinity changes [29]. Petersen et al. also observed that a salt concentration increase of 10% above ambient levels destroyed both morphology and physiology of corals [30]. It was concluded that the dual confluence of increased salinity and polyphosphate addition (to reduce scaling and fouling in RO membrane) had a greater influence on all the sample sizes of corals tested.

There were exceptions where brine disposal had negligible impacts, which have been observed on the marine flora and fauna species [31]. These only occurred where abundant currents and choppy waters were present, i.e., Australia. Studies have also point to the alleviation of detrimental impacts brought about by brine disposal by suggesting the long-term use of multiport diffusers [32, 33].

Brine with temperatures higher than ambient seawater temperature by 10–12° may have several harmful effects on marine fauna and flora. The toxicity effects brought about by metals and chemicals amplify significantly with temperature [34, 35]. Furthermore, various heavy metals like copper and nickel (found in alloys of heat exchangers) inadvertently become part of the brine stream when corrosion of heat exchangers occurs during desalination. In 2016, Alshahri investigated the concentrations of heavy metals in disposed brines of Persian Gulf desalination plants [36]. Sand and sediments near the Persian Gulf were found to contain copper, iron, and chromium, and their concentrations were higher than that in shale due to anthropogenic activities. Likewise, research on the Al-Khafji coastal sediments point to high levels of copper at the periphery of the coastline located north, plausibly due to brine disposal from nearby coastal desalination plants [37]. When evaporation ponds are unlined, the soil quality declines due to ionic replacements of calcium with sodium ions [21, 24], and underground aquifers are contaminated when brine is discarded in such areas [38, 39].

3. Common strategies for brine solution disposal

Large amount of brine is generated after the desalination process that many disposal options are implementing currently. According to previous sections, it has been mentioned that surface water discharge, deep well injection, land application, evaporation ponds, and conventional crystallizers are all considered as traditional methods for brine disposal. The quality and volume of the concentrate, as well as the physical and geographical locations of the output point of the concentrate, are the factors affecting the disposal options of concentrate. Also, to select an appropriate disposal option, it is necessary to consider the economic aspects, feasibility, general acceptance, authorization of the option, the availability of disposal site, and also the feasibility of facility development. One of the most important factors to be considered before selecting an option is the cost of brine disposal which is a barrier to the extended use of this process [40, 41]. Between 5 and 33% of the total cost of the process is usually related to brine disposal, and it depends on the features and volume of the brine, disposal option, and the level of brine treatment before being disposed. The details of different conventional brine disposal options are described in the following sections.

3.1 Surface water discharge

Discharging the brine directly to the open aquatic environment like lakes, rivers, bays, and oceans is considered as surface water discharge. The brine, after being transferred to a disposal site, sheds into the desired aqueous medium through a special structure. Most SW desalination plants are designed and perform based on this method (N90% of world SW plants). On the contrary, inland brackish water desalination plants are more limited based on that inland water bodies can be used as water sources because of their high quality. On this basis, the discharge can only be done when the composition of the brine is consistent with the receiving water body and suitable for discharging on that point [42, 43]. As previously mentioned, because of the higher-than-usual salinity of the brine or its ingredient pollutants that do not exist in the receiving water body, it can be detrimental to the marine environment. Appropriate restrictions and measures can make the brine disposal in surface water as a sustainable method for SW desalination plants [44]. For example, dilution of the brine using the municipal wastewater or regular SW before discharging to the marine environment is a kind of measure to decrease the salinity level of the discharging brine [45]. Research has shown that if dilution and rapid mixing are used with caution to decrease the concentration of the brine, it has a negligible adverse impact [46]. The cost range of this disposal method is from US$0.05 to US$0.30 based on 1 m^3 of brine rejected.

3.2 Discharge to the sewage system

A brine disposal method in which the brine is discharged into the nearby sewage collection system is called sewer discharge. Most of the small-scale BW desalination plants are using this method of discharge as the high TDS content available in the brine can potentially have a negative impact on the receiving wastewater treatment plant (WWTP) [47]. Generally, a high level of TDS content in influent which its concentration exceeds 3000 mg/L can inhibit the biological treatment process in a WWTP as the salinity is very high in this situation [48]. Given the concentration of SW brine TDS can exceed 55,000 mg/L, the daily capacity of WWTP must be at least 20 times higher than the volume of brine discharge in order to maintain the restricted range of influent TDS concentration lower than 3000 mg/L. Furthermore, a very high salinity of final wastewater effluent can have issues related to the environment and regulations during the final disposal process. In addition, the probable existence of heavy metal traces in the brine may require some measures like pH neutralization or other regulating procedures before the main treatment process as a basic pretreatment. Such measures guarantee the substructure of the treatment process and also the quality of the final wastewater effluent [49]. Accordingly, sewer discharge is broadly used by BW desalination plants and is not consistent with SW desalination processes. The cost range of this disposal method is from US$0.32 to US$0.66 based on 1 m^3 of brine rejected [50].

3.3 Deep well injection

The procedure in which the brine is injected into a deep aquifer existing beneath the groundwater layers is called deep well injection. The most important matter must be considered before the injection is to ensure that wastes are not leaked to other locations, and the capacity of the target aquifer must be consistent with the plant life, and also this aquifer should be hydraulically isolated from surrounding porous media. This method is most suitable for disposal of the brines without monovalent cations and heavy metals as it can prevent the precipitation of such materials before disposal [40]. This method is usually used for municipal, industrial, and liquid hazardous wastes and requires a suitable geological circumstance [51]. It is necessary to evaluate the geological conditions of the injection site in detail and specify the depth and exact location of suitable porous media before drilling an injection well [52]. Deep well injection has the highest rate of capital cost among other disposal options. Finding an appropriate well site, corrosion and leakage of the wastes into the well casing, and subsequent groundwater contamination are the most important challenges regarding this disposal option [52]. Accordingly, this method is only used when there is no suitable alternative.

3.4 Application for land

In some situations, the brine can be used for vegetation such as parks, golf courses, and lawn irrigation, and land application of the brine can be suitable to reuse water for these purposes. This process can also provide the required nutrients for the plants. Different factors are affecting the selection of this option such as the existence and price of the land, expenses related to water dilution, cost of the irrigation equipment, infiltration rates, irrigation importance, salinity tolerance interval for desired vegetation, and groundwater quality regulations [53]. Based on the Food and Agriculture Organization (FAO) of the United Nations [54], the regulated concentration limits of Ca, Mg, and Na ions to irrigate the general crops are 400, 60, and 900 mg L^{-1}, respectively. Although brine may have negative impacts on the soil and groundwater when disposed in the land, the reuse of brine originating from livestock wastewater in the agricultural application as a liquid fertilizer is proposed by Yoon et al. [6]. This recommendation is only applied when there are no pathogenic microbes in the sample. The adverse impact on underground aquifers has been found by Mohamed et al. when the brine is directly disposed in some lands with permeable soil containing a low clay content and organic matter [38]. Increasing the concentration of the brine can reduce the permeability of the soil and consequently can reduce the crop yield. Therefore, based on the salinity tolerance interval of the crop for different ions, the brine must be diluted with freshwater to an acceptable range.

3.5 Evaporation ponds

In evaporation ponds, brine is directly under the sunlight and slowly evaporates in shallow, arrayed earthen basins. When the freshwater available in the brine has evaporated, the solutes in the brine are precipitated and then periodically removed from the site. The evaporation ponds are broadly used in some locations where the temperature and dryness are high enough to evaporate water at the desired time [55]. Since there are some critical concerns regarding groundwater pollution, this method must be appropriately designed and performed. In general, based on environmental regulations, the evaporation ponds are forced to be isolated from underlying aquifers using the impermeable coating. In some situations with high levels of rare metals, a pond with double liner has to be constructed. In addition, when the ponds are not laminated or the point liner is damaged, some brine may infiltrate the aquifer beneath the pond and reduce its water quality [56]. Some factors are affecting the selection of this disposal option such as the existence and price of the land, climate circumstances, and underlying groundwater quality. The cost range of this disposal method is from US$3.28 to US$10.04 based on 1 m^3 of brine rejected and is the most expensive disposal option [43].

3.6 Conventional crystallizers

The brine contains some metals that can be recovered as an attractive solution to avoid disposal problems with additional economic advantages. Thus, it is a critical challenge for researchers and industrial activists to manage brine concentrate regarding its beneficial resource recovery. During the recovery process, if rare and valuable components are found, it can be considered as a double objective. This achievement can enhance the overall economic efficiency of the treatment process by reducing the adverse impacts of RO concentrate disposal on the environment. In the last stage of the brine disposal process, some operations are performed for brine crystallization. Compared to other disposal methods such as evaporation pond and deep well injection, the brine crystallizer process is expensive [40]. This process is only executable where deep well injection treatment is expensive, evaporation ponds cost a lot for its instruction, and the rate of evaporation is low. In Israel, a brine discharge from RO plant is fed to a series of evaporation ponds after mixing with seawater and then pump to a salt processing plant [57]. The evaporation and crystallization steps for salt recovering from RO brine have been evaluated by Mohammadesmaeili et al. using lime softening in several stages of the evaporation crystallization processes [58, 59]. They found that magnesium hydroxide, calcites, and $CaSO_4$ with the purity of 51–58, 95, and 92%, respectively, were produced after lime soda treatment. Zero liquid discharge achievement by combining RO with evaporation and crystallization was noted by Seigworth et al. [60]. Therefore, while additional study is required to evaluate the economic aspects of salt production, using these methods for salt recovery is emphasized as a sustainable option. Ahmed et al. used the patented SAL-PROC (Geo-Processors Inc., USA) processes to determine the sustainability of recovered salts from RO retentate of a desalination plant by consecutive extraction of salts in the form of liquid, crystalline, and slurry [61]. They specified that sodium chloride, $CaCO_3$, sodium sulfate (Na_2SO_4), and calcium chloride were the most likely recoverable products that can provide the potential cost-effectiveness of the desalination processes. It has been found by Ahmad and Williams in another study that 45 million tons of salts are recovered yearly in the USA and chemical industries use about 70% of this amount [62].

4. Common technologies for brine treatment

Brine contaminant is one of the main concerns due to hazard creation in the environment. Therefore, it is necessary to remove them before safe disposal in the beneficial use of reclaimed brine solution or open water bodies. Biological processes, oxidation, and chemical precipitation coagulation can be used for the treatment of brine.

4.1 Chemical precipitation process

Chemical softening has been broadly utilized for the brine treatment with lime softeners. High removal of scale-forming ions is the main advantage of utilizing the chemical softening framework for concentrate treatment. Be that as it may, the confinement of the method is the era of slime which needs additional care for legitimate management. Lime treatment method was utilized by Kolluri for removing the silica from the brine which results in the removal of silica content by 53–76% [63]. It was noted that lime treatment was exceptionally successful to remove silica from high silica brine and found that no silica removal happened until the lime dose surpassed the lime equivalent of the alkalinity [64]. Gabelich et al. have shown that the Ca substance was removed as $CaCO_3$ after elimination of silica and metals (such as Ba) by coprecipitation with $Mg(OH)_2$ [65]. In the absence of alkalinity, sodium bicarbonate was included in the brine to precipitate the Ca as $CaCO_3$. Be that as it may, the nearness of harmful overwhelming metal particles and developing natural contaminants in brine arrangement might prevent the immaculateness of the target compounds' precipitation, and this still ought to be completely investigated.

4.2 Coagulation

Coagulation may be a basic physicochemical and commonly connected handle for organics' evacuation from water and wastewater. The components included in coagulation are charge neutralization and adsorption of organics on the metal hydroxide [66, 67]. The type and dosage of coagulant used and the characteristics of the feedwater impact the efficiency of organics' removal [68]. Coagulation has not been considered broadly for utilizing in brine treatment. This can be since brine contains an essentially high concentration of salts. Given the straightforwardness of the method, this method has been examined for eliminating the natural component from high saltiness brine arrangement. Umar et al. explored coagulation utilizing two aluminum-based [alum and aluminum chlorohydrate (ACH)] and two ferric-based coagulants [ferric chloride ($FeCl_3$) and ferric sulfate ($Fe_2(SO_4)_3$)] for treatment of high saltiness brine concentrate (EC of 23 mS cm^{-1}) and showed that at 1 mM dose, the DOC elimination for the two ferric-based coagulants was comparable (40–43%) while that for ACH was extraordinarily lower (14%) than for alum (23%) treatment [69]. Dialynas et al. examined the execution of alum and $FeCl_3$ coagulation for the concentrate gotten from a plant and showed that the DOC elimination was 42% for alum (beginning DOC, 8.5 mg L^{-1}) and 52% for $FeCl_3$ (starting DOC, 12.3 mg L^{-1}), with an ideal dose of 2 mM as Al^{3+} and 0.4 mM as Fe^{3+}, individually [70]. This shows that iron-based coagulants are more effective than aluminum-based ones for brine treatment. Employing a higher concentration of $FeCl_3$ (1 mM Fe^{3+}) than Dialynas et al. [70], a lower evacuation of DOC of 26.4% (introductory concentration, 18 mg L^{-1}) from brine was detailed by Zhou et al. Comstock et al. examined the impact of $Fe_2(SO_4)_3$ doses (1.79, 4.48, 8.95 mM Fe^{3+}) for brine treatment (introductory DOC, 13 mg L^{-1}) gotten from a civil drinking water treatment plant [71, 72].

In spite of the fact that the initial DOC concentration of the brine arrangement was comparable with that of Dialynas et al., they utilized an essentially higher dose of coagulant (8.95 mM Fe^{3+}) for a comparative degree of DOC evacuation (58%) [70]. This was conceivably due to the diverse water source and in this way diverse organics' characteristics. Compared to the already specified postulates, Bagastyo et al. found a lower expulsion of DOC of 52% for a starting DOC concentration of 42 mg L^{-1} and 25% for a starting DOC concentration of 62 mg L^{-1}, with alum coagulation [73]. But, $FeCl_3$ gave a comparable elimination of DOC of 34% and 38%

for both tests. They have shown that the ideal measurement for both coagulants was 1.5 mM at pH 5. Researchers have shown that alum coagulation might eliminate up to 42% of DOC, while $FeCl_3$ might eliminate 52% at the same dosages [70]. Another research by Bagastyo et al. [59] has shown that $FeCl_3$ coagulation might eliminate 79% of color, 34% of DOC, and 49% of COD. The lower elimination of DOC and COD was due to the existence of a high percentage of low to medium molecular weight (MW) compounds within the brine solution. This was since this method was incapable to expel solvent natural compounds with a low MW because it might primarily expel high MW organics [74].

4.3 Electrocoagulation

Electrochemical treatment is an effective treatment innovation for the treatment of tall saltiness water because it guarantees a fabulous electric conductivity that might diminish the vitality utilization. This handle incorporates an electrolytic reactor with anodes (aluminum, press, or stainless steel) and a division tank in which feedwater is passed through a reactor, and coagulation/flocculation happens with the metal dissolved from the cathodes [75]. The metal anode dissolution is going with by hydrogen gas bubble arrangement at the cathode, coming about in capture of the flocs, and this at that point causes buoyancy of the suspended solids, inevitably eliminating the contaminants. The preferences of this method incorporate less slime generation than an ordinary coagulation method. The challenge of this method is the high operation and support costs related to anode substitution, high vitality utilization, and restricted full-scale plant involvement. Subramani et al. explored the impact of electrocoagulation for the treatment of brine, and it is known that this method was exceptionally productive in evacuating Ba, Ca, Mg, strontium (Sr), and silica with more than 90% expulsion [10]. Another research by Cave and Wang utilized electrocoagulation as a pretreatment step for the brine preparation for anticipating silica fouling and has shown that 80% of the silica was eliminated at a current concentration of 0.5 A and water-powered maintenance time of 30 min [76].

4.4 Oxidation-based technologies

4.4.1 Ozonation

Ozonation has been broadly utilized for water and wastewater treatment, especially for the corruption and advancement of biodegradability of the natural compounds [77, 78]. The organics are oxidized either through a coordinate response with molecular ozone (O_3) which is profoundly particular or backhanded responses with free radicals ($HO^{\cdot}$) [79, 80]. This method was developed to treat brine either alone or in combination with other methods. Stand-alone O_3 was used by Lee et al. [81] and Zhang et al. [82] to degrade the organic content of brine solution with similar initial TOC and COD concentrations for both samples of 18 mg L^{-1} and 60–65 mg L^{-1}, respectively. The impact of ozonation on the degradation of organic compounds from the brine solution was investigated by Zhou et al. [77]. It was shown that the elimination efficiencies of COD, DOC, and color were 14%, 22%, and 90%, respectively. Lee et al. [81] conducted a series of experiments to investigate the impact of reducing the organic compounds by using an ozonation method and showed that the TOC was reduced by 25% after 20 min of ozonation (at a 10 mg L^{-1} concentration). They also found that the enhancement of TOC reduction was less than 2% after increasing the dosage from 6 to 10 mg L^{-1}, highlighting that it was difficult to remove the remaining organics using decomposition via O_3. Lee et al. [81] have shown that the large MW organics were converted to low MW organics and explained the difference in the distribution of the organics' fractions after brine treatment. Furthermore, they found that organics with an MW of 10–100 kDa were mostly reduced (36–72%) after 10 min, and the removal efficiency is improved by 20% when the contact time increased from 10 to 20 min. It was also shown that it is least likely to remove the organic compounds with a MW of >100 kDa. Most probably this was due to the surface trimming effect of O_3 which results in breakage of some of the bonds on the surface but incomplete molecule disintegration, resulting in bulk organic content intact [83]. By utilizing the O_3 process of brine treatment, Le et al. analyzed the process before using the biological activated carbon (BAC) process. They showed that only 5% of TOC will be removed if we use O_3 alone with 3.0 mg O_3 L^{-1} with 10 min contact time. It was also observed that BAC treatment (60 min contact time) is capable of giving better removal of organics than O_3 alone (6.0 mg O_3 L^{-1} with 20 min contact time), and combined O_3-BAC processes removed 88.7% of TOC and 69.8% of COD. This is mainly due to the usage of O_3 which results in an enhancement of the biodegradability of the organic compounds. This is ascribed to the effective breakdown of high MW aromatic compounds to low MW organics which is followed by appropriate biodegradation of the residual organics with the BAC system. As shown by Tambo and Kamei, the ratio of TOC/UVA254 is utilized to assess the performance of biodegradability, and they found that by 20 min of ozonation the ratio was enhanced from 35 to 107 [84]. This also approves the helpfulness of O_3 in breaking down the large MW organics and enhancing the biodegradability of the residual treated organics. It was shown that after 30 min ozonation treatment of the brine the COD removal was 19–25% [78]. The lower reduction was mainly due to the insufficient reaction of O_3 with the by-products. The combination of processes such as UVA/titanium dioxide (TiO_2), ultraviolet A/hydrogen peroxide (UVA/H_2O_2), and O_3 was investigated, and they showed that the combined method was not effective at enhancing the DOC reduction compared to ozonation alone [46].

4.4.2 UV/H_2O_2 process

UV/H_2O_2 technology is of great interest to treat the brine solution [46]. Researchers found that treatment by UV/H_2O_2 can effectively remove the organic compounds over a wide range of MW, and the low MW organics react slower than large MW compounds due to the fact that smaller organics are less aromatic in nature and contained lower molar absorptivities, thus having a small number of reaction sites available to react with $HO^{\cdot}$ [85]. Application of UV/TiO_2 for the removal of the organic load from the brine concentrates was investigated in several research projects. Westerhoff et al. investigated the impact of an ultraviolet (UVC)/H_2O_2 method to treat the brine concentrate with an initial DOC concentration of 40 mg L^{-1} and showed that 40% of DOC was eliminated by utilizing the 10 mM H_2O_2 at pH 4 [86]. A research conducted by Zhou et al. demonstrated poor performance of UVA/H_2O_2 to treat the brine where 10 mM H_2O_2 dosage at pH 4 could only reduce the 2.3 ± 2.8% of DOC [77]. This poor removal efficiency is mainly ascribed to the better molar absorption coefficient of H_2O_2 for UVC at 254 nm than the UVA at 360 nm [77, 87]. It was shown that DOC reduction by UVC/6 mM H_2O_2 treatment is greater than VUV/2 mM H_2O_2 and UVC/2 mM H_2O_2 treatment [77, 87]. It was also found that UVA254 has higher reduction which is mainly due to the breakdown of the chromophoric and conjugated structure of the organic compounds. Bagastyo et al. [73] analyzed the impact of the UVC/H_2O_2 method in treating the variety of brine samples collected from two different wastewater treatment plants (DOC of 42 and 62 mgL^{-1}) and showed that the DOC removal efficiency is comparable (38% and 40%) as compared to the results provided by Westerhoff et al. [86]. Furthermore, it was shown by Bagastyo et al. that the removal of dissolved organic nitrogen for both samples was insufficient (32 and 27%) [73]. This was mainly due to the high proportion of low MW organic compounds present in the brine sample which were positively or neutrally charged compounds with low reactivity to oxidation by the UVC/H_2O_2 process. The UVC/H_2O_2 method was investigated by Umar et al. in treating one moderate (EC ~8 mS cm^{-1}) and two high-salinity (EC ~23 mS cm^{-1}) municipal wastewater samples with different inorganic and organic characteristics [12]. It was shown that the difference in the reduction of DOC (26–38%) and COD (25–37%) over the tested saline conditions is very small. This indicates that the brine solution salinity did not have a substantial influence on the treatment of UVC/H_2O_2. It was also found that UVA254 and color reduction were substantially higher than for COD and DOC for all samples, indicating the larger humic compound breakdown. However, only few researchers have analyzed the impact of level of salinity on the performance of UVC/H_2O_2 for organics removal. TiO_2 suspension by UVA irradiation was investigated by Dialynas et al. [70] in treating the brine solution and showed DOC reduction after a 60-min reaction. This was mainly due to the enhancement of opacity caused by the suspension of catalyst. The impact of ozonation in treating the brine with/without UV/H_2O_2 was also explored and showed that O_3 only is capable for removing only 22% of the DOC; however, the combination of O_3 with TiO_2 and UV could improve the efficiency of removal by 52% [88].

4.5 Biological processes

One of the important factors affecting the efficiency of biological processes is the existence of high salinity in wastewaters. This is mainly due to the high-salinity concentration which results in unbalanced osmotic stress across the microbial cell, resulting in systems failure [89]. In addition, due to the existence of bio-refractory organic compounds in the brine solution, the biological processes are not very effective [77]. Häyrynen et al. demonstrated that the use of bioreactors to remove the nutrients and the existence of heavy metals such as chromium and Cu in the feed-water inhibit the efficiency of the nitrifying bacteria [90]. Ersever has conducted a research to investigate the several biological processes to remove the sulfate, nitrate, and NH_3 from the brine [5]. It was shown that at a temperature of 35°C, pH of 8·0, and C/N of 1·8:1, the maximum denitrification rates can be achieved. By considering the nitrogen as a minor pollutant in the waste stream, Ersever et al. investigated the impact of a fluidized bioactive absorber reactor method in removing nitrogen compounds from RO brines and showed that this technique is very effective to remove the nitrogen (90%) from RO concentrates [91]. As shown by Dialynas et al., the membrane bioreactor can effectively remove the organics from the RO concentrate [70]. It was shown that 90% of the organics from RO concentrate can be removed by activated carbon due to the adsorption of medium/small MW organics by the pores of the activated carbon [70, 77].

Ng et al. [92] and Lee et al. [13] investigated the removal of organic content using BAC columns. It was shown by Ng et al. that 39.6% of COD and 25% of TOC from brine solution with an empty bed contact time (EBCT) of 40 min can be removed [92]. So far most of the researches focused on using the BAC method for treating the brine with a TDS level of less than 2000 mg L^{-1} [40]. A research by Vallero et al. demonstrated that biodegradation of methanol in a non-adapted granular inoculum sludge system can be completely inhibited by high NaCl concentration (25 gL^{-1}) [89]. Shi et al. demonstrated that a TDS concentration more than 14.92 g L^{-1} considerably reduced the removal of COD by the upflow anaerobic sludge blanket and utilized an ion exchange resin pretreatment to eliminate the salt concentration present in the wastewater while holding the most portion of organics in the wastewater and showed that 22% of the COD was eliminated together with the elimination of 80% of TDS [93]. Lu et al. investigated the impact of BAC with high salinity (TDS levels of 10 g L^{-1}) for treating the brine concentrate. It was shown that the reduction efficiency of COD and DOC was approximately 50 and 60%, respectively [94]. This indicates the effectiveness of biological treatment for activated sludge to be acclimated to high-salinity environments.

5. Membrane technologies

5.1 Forward osmosis

Contrary to the energetically intensive hydraulic pressure-driven reverse osmosis (RO), forward osmosis (FO) drives water through membranes due to osmotic pressure differences that are inherently present in the system. The driving force across the membrane is attributed to the differences in salt concentration and creates the osmotic pressure gradient [95]. Water moves from the feed (low salt concentration) to the draw solution (high salt concentration) [95]. There have been many multifarious studies conducted on FO applications, for instance, minimizing the leachate from sanitary landfills [95], reducing salt content in draw solutions [95–98], identification of fruit juice concentrations [95], provision of emergency water supply [99], reducing RO [100] and anaerobic digester concentrate [101], volume production, and lastly, treatment of wastewater with high salt content in petrochemical and fracking industries [101, 102]. A study conducted by Wang and Ng employed draw solution containing 5–6 M of fructose and treated reverse osmosis concentrates (ROC) consisting approximately 1.5 M of NaCl—passing through a FO membrane with a cellulose acetate matrix [100]. Feedwater recovery of approximately 75% was attained after 18 h; an initial flux of around 8.0 $L/m^2/h^1$ was also obtained [100]. McGinnis et al. used the draw solution consisting of NH_3/CO_2 coupled with a thin-film composite (TFC) FO membrane for the treatment of highly saline water with a total dissolved solids (TDS) concentration of approximately 75,000 mg/L [102]. Water flux obtained averaged out to be 2.5 $L/m^2/h^1$ and a recovery of about 65% was attained. TDS concentrations from the FO process were in compliance with discharge standards of under 300 mg/L. Moreover, FO consumed about 40% less electrical energy than the forced circulation mechanical vapor compression (MVC) systems applied conventionally [102]. It was reported that by utilizing the draw solution consisting of 26% NaCl and a cellulose triacetate (CTA) membrane for produced water treatment, an average flux of 6 $L/m^2/h^1$ at recovery rate of 50% was recorded [101]. In Hancock's study, FO process was utilized to treat produced water with TDS ranging from 70,000 to 225,000 mg/L. Data obtained showed that recovery averaged to be 60% with flux hovering around 3 $L/m^2/h^1$ [103]. The entire configuration in Hancock's study was able to comply with the United States Environmental Protection Agency's (USEPA) effluent standard—inclusive of heavy metals, radioactive matter, halides such as Cl^-, and TDS. Boron concentrations of less than 0.05 mg/L were also met, a criterion to meet to achieve best practice in the agricultural sector [103].

The key in using FO in ROC treatment is the low energy consumption that comes with it. There is no need for external hydraulic pressure sources which are energetically intensive [95]. High TDS water exceeding 70,000 mg/L can be treated, making FO process suitable for ROC treatment. **Figure 1** shows a FO schematic employed in ROC treatment. For mining sectors, a two-stage RO was recommended for recovery of the draw solution during ROC treatment [103], compared

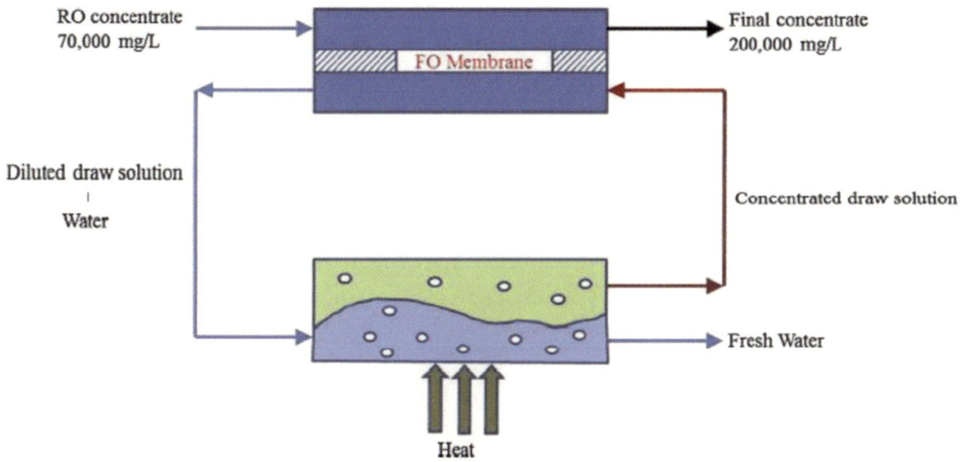

Figure 1.
Process schematic of FO process for RO concentrate treatment (adapted from [103]).

to conventional thermal-based draw solution recovery systems. Overall feedwater recoveries amounted to nearly 90% with this configuration.

Other advantages of the FO process include a lower fouling propensity of the membranes than micro-, ultra-, and nanofiltration and reverse osmosis processes, which are all pressure driven. Unlike pressure-driven membranes, the effects of membrane fouling are also more reversible in FO membranes and can be minimized by optimizing the process and hydrodynamic parameters [104–106].

Despite the multifaceted benefits, the FO system is beset with flaws. FO membranes with high efficacy, coupled with the choice of draw solutions which should be easily separable and have high osmotic pressures, are common challenges in their manufacturing. Moreover, low water flux is common in the FO process, which is contrary to the flux expected given the bulk osmotic pressure difference and permeability of membranes used. The reason for such a discrepancy is the existence of internal concentration polarization (CP) [97]. Consequently, the FO processes need to be optimized to reduce the effects of CP and damage of membrane integrity due to membrane fouling. Despite the suitability for concentrate volume. Although the application of FO has been shown to be suitable for reducing the concentrate's volume, improvements are still necessary to maximize recovery of draw solution. There are still no full-scale facilities using FO for minimizing the volume of concentrate till today [95].

5.2 Membrane crystallization

An extrapolation of membrane distillation (MD) application is membrane crystallization (MCr). MCr enables the simultaneous provision of potable water and precious crystalline salts [106]. On both sides of a hydrophobic and microporous membrane surface, it consists of a feed, which contains a solution that is nonvolatile and a distillate on the other [107]. The vapor pressure disparity between the two sides of the membrane creates a driving force that causes evaporation of volatile constituents, inclusive of water. This enables it to pass through the membrane and condense on the distillate side. The process continues until the induction of solution supersaturation and when the salt crystals nucleate. MCr systems display all the positive traits observed in the MD process, such as higher than average crystallization rates, well-controlled crystal nucleation, and growth kinetics [108]. However, compared to MD, there are limited studies on MCr present in literature. In other study, both PVDF and PP hollow fiber membranes were employed for lab-scale and semi-pilot scale MCr processes, achieving a water recovery of approximately 40% and almost 16.5 kg of NaCl salt crystals (99.9% purity) from 1 m^3 of high-saline feed solution that has a TDS concentration of nearly 250,000 mg/L [109]. Quist-Jensen et al. propound that both MD and MCr can be applied in industrial wastewater treatment containing high Na_2SO_4 content as well as direct treatment of wastewater that has not been subjected to any forms of filtration, i.e., nanofiltration [110]. The SEC and treatment costs in MCr are slightly higher than in MD, weighing in at approximately 40–75 kWh/m^3 and US$1.25/$m^3$ of freshwater produced, respectively [109, 111].

5.3 Membrane distillation

Membrane distillation (MD) relies on the fundamentals of evaporation and the separation of two or more aqueous solutions at different temperatures. A gas-liquid interface is created as volatile constituents are transferred through a microporous hydrophobic membrane [112–114]. MD occurs when there is a difference in the solution's partial pressure on both sides of the membrane [115]. If the solution's vapor pressure is higher than the condensate's vapor pressure, evaporation will occur. There are many permutations in the types of MD configurations, but for desalination purposes, direct contact membrane distillation (DCMD) is the preferred choice. In this process, an aqueous cooler distillate stream flows on one side of the hydrophobic membrane, while hot brine flows on the other. Water vapor passes through while repelling the liquid molecules due to the hydrophobic properties of the membrane [95].

When water vapor evaporates from the hot brine at the periphery of the brine-membrane interface, it diffuses through hydrophobic membrane pores which are filled with gas. The water vapor then condenses in the membrane interface at the side whereby the cooler distillate flows. By heating the feedwater, vapor pressure is increased and thus enhancing the driving gradient for vapor production [116]. Using MD alongside a crystallizer for the treatment of ROC with a conductivity of 15 mS/cm, Tun and Groth obtained an average flux of 4 $Lm^2 h^{-1}$ and an overall feedwater recovery of 95% [117]. The use of a vacuum multi-effect membrane distillation (V-MEMD) system was employed to treat concentrates of thermally desalinated seawater and improve the recovery of feedwater. **Figure 2** shows a schematic of this process. TDS in the thermally desalinated seawater's concentrate was 100,000 mg/L, and the flux of the four-stage V-MEMD system was approximately 5 $Lm^2 h^{-1}$ [118]. Although the flux from a V-MEMD system was five times lower than that of a DCMD system, the former was less energetically intensive due to the utilization of waste heat to raise the feed's temperature. In addition, having the membrane to be staged in series resulted in downstream condensation of vapor, and overall energy input is reduced because it is transferred back to the feed [118]. MD supersedes other desalination processes because operating temperatures need not exceed 70°, which is lower than the minimum temperature requirements of a conventional distillation process [119]. MD can also be retrofitted with heat sources such as renewable solar energy, geothermal energy, or waste heat sources [119]. In addition, MD efficacy is hardly affected by the CP phenomena, which enables high salt concentrations nearing saturation limits to be fed into the process. However, since MD is always associated with a low permeate flux compared to RO membrane processes [119], studies of several polymers—polypropylene (PP), polytetrafluoroethylene (PTFE), polyacrylonitrile (PAN), and polyvinylidene fluoride (PVDF)—have been conducted to circumvent this issue [120, 121]. The pivotal reason why

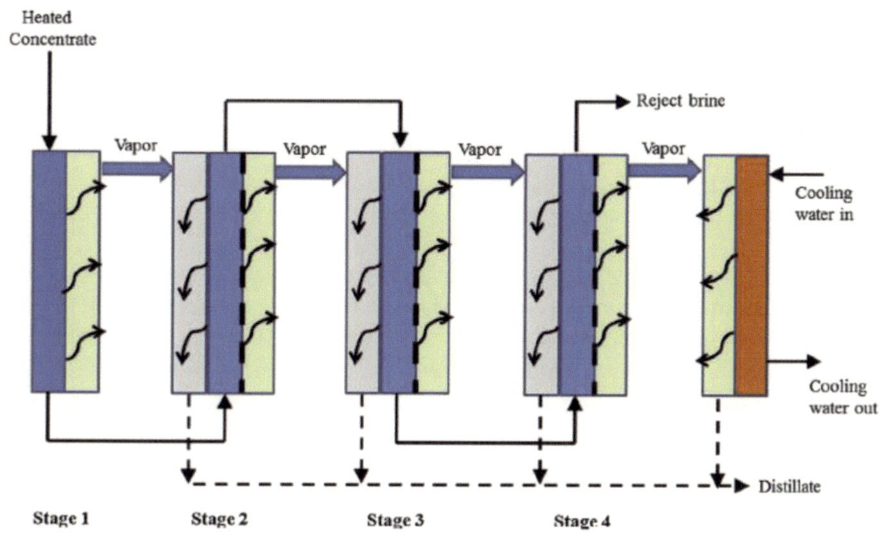

Figure 2.
Process schematic of MD process for RO concentrate treatment (adapted from [118]).

a single-layer PVDF membrane had lower decreases in membrane permeability was primarily attributed to its morphology and pore size as compared to the wall thickness of the membrane [121]. This was also in comparison to dual-layer hydrophobic-hydrophobic PVDF and dual-layer hydrophobic-hydrophilic PVDF/PAN membranes. In an alternative research, the MD polymer matrix was modified to maximize permeability by inserting carbon nanotubes. Incorporation of such nanotubes provided higher contact angles (113°), higher porosity (90%), and lower thermal conductivity [122].

6. Brine adaptation for industrial use

6.1 Brine adaptation for chlor-alkali industry

At saturated concentrations at around 300 g/L, brines containing NaCl are required to be fed into electrolytic cells for generating chlorine and sodium hydroxide. The NaCl brine must be rid of organic detritus and alkaline earth metals like strontium (Sr), barium (Br), magnesium (Mg), and calcium (Ca). Notably, Melián-Martel et al. propounded to employ a multi-effect evaporator coupled with chemical precipitation to remove Ca, SO_4^{2-}, and Mg in the treatment of seawater RO brine for the chlor-alkali process [123]. To validate the efficacy of the system, concentrated brine of approximately 8500 m^3/day from the Pozo Izquierdo desalination plant, Gran Canaria was used as the feed source to be treated. The production capacity of the said desalination plant is about 35,000 m^3/day with a conversation efficiency of 50%. An electrolyzer was used to assess the chlorine and sodium hydroxide production of the treated brine. The proposed system attained the production of 102 kilo tons/year of chlorine, 254 kilo tons/year of sodium hydroxide, and 3 kilo tons/year of hydrogen gas after treating the seawater RO brine. **Table 2** depicts the compositions of the products. A common practice to offset the cost in the chlor-alkali industry is to sell the concentrated brine to another industry. From a cost-base perspective, it is economical to produce brine from reverse osmosis concentrate than to create brine from raw seawater. Melián-Martel et al. gave a rough estimate of 2000 kWh required per ton of NaOH generated. Moreover, the 3

Cl_2	NaOH	H_2
$Cl_2 > 98\%$	NaOH—32%	>99.9%
$H_2 < 2\%$	NaCl <20 ppm	

Table 2.
Composition of end products of chlor-alkali electrolytic process [123].

kilotons/year of hydrogen gas generated could be used to generate electricity and minimize the amount of energy required. This makes the process suitable in places where energy resources are expensive or scarce. On a similar vein, brine adaptation for the chlor-alkali industry is not land intensive, which contributes to the reduction of capital costs, alleviating the strain on resources in land expensive or scarce countries [123].

Despite the positive attributes derived in brine adaptation for the chlor-alkali industry, there are problems associated with it as well. RO brine contains higher divalent cation concentration, which necessitates removal procedures involving high costs. To circumvent this problem, production of NaCl needs to occur first before generating the necessary brine of a specific matrix.

In the salt manufacturing industry, Tanaka et al. propounded that energy consumption via electrodialysis of seawater RO brine was 20% less than raw seawater [124].

6.2 HCl and NaOH production with bipolar membrane electrodialysis

The mechanism of membrane electrodialysis involves two aspects—applying a potential across the membrane to enhance mobility of ions and to limit their movement via selective membranes. To split water into its hydrogen ions (H^+) and hydroxide ions (OH^-), the application of electrodialysis is done in conjunction with a bipolar membrane—hence the name bipolar membrane electrodialysis (BMED) [125]. The combination of hydroxide ions and cations and combination of protons and anions lead to production of acid and base respectively. **Figure 3** shows the diffusion of ions across the membrane in a BMED system. Badruzzaman et al. propound the employment of BMED as the final step for generating hydrochloric acid and sodium hydroxide from high-salinity solutions after undergoing sequential treatment steps of using membrane bioreactors, coupled with reverse osmosis and softening via calcium hydroxide. The entire system is known as an integrated membrane system (IMS) [126]. In Badruzzaman et al.'s study, the salinity of the feed was about 3 g/L, which deviates greatly from seawater or brine water salinity. Research also validates that electrodialysis accumulates major positive cations and negative anions in the acid and base chamber, respectively. Water after the treatment process can be used directly as product water or subject to further reverse osmosis treatment for further purification. The authors compared the capital costs for IMS implementation. The different permutations are as follows: the first involved a sequential MBR and RO followed by an evaporation pond for disposal. The second permutation revolved around a zero liquid discharge thermal process via a concentrator and a crystallizer. The capital and yearly operational and maintenance (O&M) costs for first option are approximately \$1.65 and \$0.41/m^3, while that of the second option are \$0.50 and \$0.80/m^3, respectively. IMS, on the other hand, only requires \$0.43 and \$0.25/m^3. These cost values for the IMS do not factor in the profit margin of selling the acid and base produced, which are chemicals of high demand, alongside the cost of recovered water. The O&M costs can be alleviated by

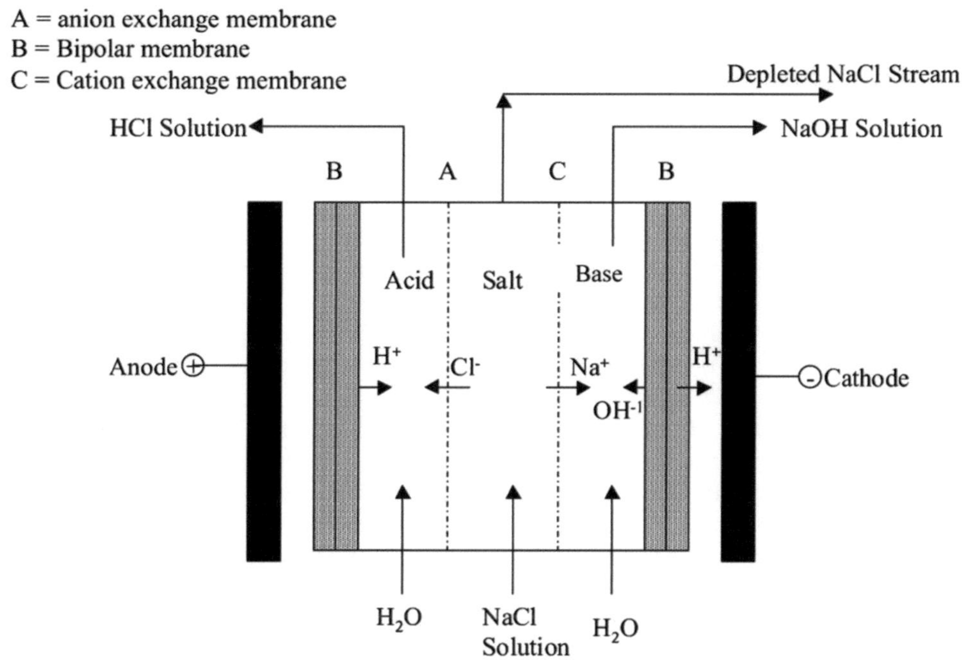

Figure 3.
Schematic of bipolar membrane electrodialysis (BMED) system operating principle [126].

up to \$0.1/m^3. Unfortunately at present, BMED is not applied at an industrial level because the electrolytic cells applied are not as commercially established.

7. Metal recovery

Sixty elements from the periodic table are usually existed in seawater that some of them are rare and more valuable. Precious metals are valuable components in seawater, and their recovery from the rejected brine has long been considered for their advantages due to the relatively high levels in retentate brine. Based on several physicochemical, economic, and technical aspects, Dirach et al. suggested a protocol to recover elements of interest from concentrate. This process uses evaporation to increase the concentration of solution up to about 200 g/L before the first recovery step to extract phosphorus precipitation using an alum blend of iron sulfate and aluminum sulfate [127]. Then a liquid-liquid extraction is applied by adding HCl to recover cesium from the solution. Another liquid-liquid extraction will be performed using an organic phase, consists of three different acids for indium recovery. A countercurrent process with 15 stages is essential to have an effective separation. Indium with a purity of 97.4% and also gallium with a purity of 99.8% will be recovered as a result of this process. Rubidium is then extracted by means of cation exchange resins. Potassium and rubidium are the first and second most attracted elements, respectively. To separate these two elements from the solution, it is necessary to increase the purification. In the next step, germanium will be undertaken which is crystalized to form germanium dioxide (GeO_2). The crystalized form of germanium is then exposed to gaseous HCl to be oxidized and then will be reduced to pure germanium by roasting in a reducing atmosphere of H_2. The main components contributing the remaining solution are mainly magnesium, potassium, and NaCl. The solubility differences between these compounds are used to separate them from each other. The potential of recovering valuable metals (uranium, rubidium, cesium, lithium) from RO brine related to a plant in El Prat de Llobregat, Spain, using a number of sorbents, has been studied by Peterskovà et al. [128]. It was concluded that the best sorbent for both cesium and rubidium between all tested ones was hexacyanoferrate-based extractant Cs-Treat, while all tested sorbents were effective enough at sorbing lithium. The resin containing phosphonic and sulfonic groups has the highest tendency for uranium(VI) sorption. Even more importantly, though, the results showed that the salinity is a crucial factor affecting cesium sorption affinity onto Cs-Treat. Single-metal systems are moderately different from bimetallic systems based on the sorption results, which is due to the independency of sorption capacity from co-ion effect, and accordingly Cs-Treat was highly selective for cesium and rubidium [125]. By utilizing metal recovery, novel and abundant sources of many valuable and rare metals can be provided from all over the world which can greatly increase its potential profitability [125].

As an example, in many countries where the conventional process is not available to produce non-carbon energies, uranium recovery would provide a non-carbon source of energy. In terms of environmental prospective, the impacts of metal recovery are much lower than mining, although the new technologies involved in metal recovery are still far from ideal conditions and require more improvement to be competitive with traditional process. Accordingly, this technology needs more attention and research to increase productivity and improve performance in metal recovery process and also adequately develop to be exploited on an industrial scale [125].

Author details

Reza Katal[1], Teo Ying Shen[1], Iman Jafari[1], Saeid Masudy-Panah[2] and Mohammad Hossein Davood Abadi Farahani[3]*

1 Department of Civil and Environmental Engineering, National University of Singapore, Singapore

2 Low Energy Electronic System (LEES), Singapore-MIT Alliance for Research and Technology (SMART), Singapore

3 Research and Development Department, SEPPURE, Singapore

*Address all correspondence to: farahani@seppure.com

References

[1] Snyder SA. Occurrence, treatment, and toxicological relevance of EDCs and pharmaceuticals in water. Ozone Science and Engineering. 2008;**30**:65-69. DOI: 10.1080/01919510701799278

[2] Solley D, Gronow C, Tait S, Bates J, Buchanan A. Managing the reverse osmosis concentrate from the western corridor recycled water scheme. Water Practice Technology. 2010;**5**:1-8. DOI: 10.2166/wpt.2010.018

[3] Ahmed M, Shayya WH, Hoey D, Mahendran A, Morris R, Al-Handaly J. Use of evaporation ponds for brine disposal in desalination plants. Desalination. 2000;**130**:155-168. DOI: 10.1016/S0011-9164(00)00083-7

[4] Van der Bruggen B, Lejon L, Vandecasteele C. Reuse, treatment, and discharge of the concentrate of pressure-driven membrane processes. Environmental Science & Technology. 2003;**37**:3733-3738. DOI: 10.1021/es0201754

[5] Ersever I, Ravindran V, Pirbazari M.Biological denitrification of reverse osmosis brine concentrates: I. Batch reactor and chemostat studies. Journal of Environmental Engineering and Science. 2007;**6**:503-518. DOI: 10.1139/S07-021

[6] Yoon Y, Ok Y-S, Kim D-Y, Kim J-G. Agricultural recycling of the by-product concentrate of livestock wastewater treatment plant processed with VSEP RO and bio-ceramic SBR. Water Science and Technology. 2004;**49**:405-412. DOI: 10.2166/wst.2004.0781

[7] Kumar M, Badruzzaman M, Adham S, Oppenheimer J. Beneficial phosphate recovery from reverse osmosis (RO) concentrate of an integrated membrane system using polymeric ligand exchanger (PLE). Water Research. 2007;**41**:2211-2219. DOI: 10.1016/j.watres.2007.01.042

[8] Gomes AC, Gonçalves IC, de Pinho MN, Porter JJ. Integrated nanofiltration and upflow anaerobic sludge blanket treatment of textile wastewater for in-plant reuse. Water Environment Research. 2007;**79**:498-506. DOI: 10.2175/106143007X156844

[9] Subramani A, Schlicher R, Long J, Yu J, Lehman S, Jacangelo JG. Recovery optimization of membrane processes for treatment of produced water with high silica content. Desalination and Water Treatment. 2011;**36**:297-309. DOI: 10.5004/dwt.2011.2604

[10] Subramani A, Cryer E, Liu L, Lehman S, Ning RY, Jacangelo JG. Impact of intermediate concentrate softening on feed water recovery of reverse osmosis process during treatment of mining contaminated groundwater. Separation and Purification Technology. 2012;**88**:138-145. DOI: 10.1016/j.seppur.2011.12.010

[11] Randall DG, Nathoo J, Lewis AE. A case study for treating a reverse osmosis brine using eutectic freeze crystallization—Approaching a zero waste process. Desalination. 2011;**266**:256-262. DOI: 10.1016/j.desal.2010.08.034

[12] Umar M, Roddick F, Fan L. Assessing the potential of a UV-based AOP for treating high-salinity municipal wastewater reverse osmosis concentrate. Water Science and Technology. 2013;**68**:1994-1999. DOI: 10.2166/wst.2013.417

[13] Lee LY, Ng HY, Ong SL, Tao G, Kekre K, Viswanath B, et al. Integrated pretreatment with capacitive deionization for reverse osmosis reject recovery from water reclamation plant.

Water Research. 2009;**43**:4769-4777. DOI: 10.1016/j.watres.2009.08.006

[14] Ning RY, Tarquin A, Trzcinski M, Patwardhan G. Recovery optimization of RO concentrate from desert wells. Desalination. 2006;**201**:315-322. DOI: 10.1016/j.desal.2006.06.006

[15] Panagopoulos A, Haralambous K-J, Loizidou M. Desalination brine disposal methods and treatment technologies—A review. Science of the Total Environment. 2019;**693**:133545. DOI: 10.1016/j.scitotenv.2019.07.351

[16] Oren Y, Korngold E, Daltrophe N, Messalem R, Volkman Y, Aronov L, et al. Pilot studies on high recovery BWRO-EDR for near zero liquid discharge approach. Desalination. 2010;**261**:321-330. DOI: 10.1016/j.desal.2010.06.010

[17] Gude G. Emerging Technologies for Sustainable Desalination Handbook. Amsterdam: Elsevier; 2018. DOI: 10.1016/C2017-0-03562-0

[18] Melián-Martel N, Sadhwani Alonso JJ, Pérez Báez SO. Reuse and management of brine in sustainable SWRO desalination plants. Desalination and Water Treatment. 2013;**51**:560-566. DOI: 10.1080/19443994.2012.713567

[19] Kayvani Fard A, Rhadfi T, Khraisheh M, Atieh MA, Khraisheh M, Hilal N. Reducing flux decline and fouling of direct contact membrane distillation by utilizing thermal brine from MSF desalination plant. Desalination. 2016;**379**:172-181. DOI: 10.1016/j.desal.2015.11.004

[20] Lior N, Kim D. Quantitative sustainability analysis of water desalination—A didactic example for reverse osmosis. Desalination. 2018;**431**:157-170. DOI: 10.1016/j.desal.2017.12.061

[21] Heck N, Paytan A, Potts DC, Haddad B. Predictors of local support for a seawater desalination plant in a small coastal community. Environmental Science & Policy. 2016;**66**:101-111. DOI: 10.1016/j.envsci.2016.08.009

[22] Frank H, Rahav E, Bar-Zeev E. Short-term effects of SWRO desalination brine on benthic heterotrophic microbial communities. Desalination. 2017;**417**:52-59. DOI: 10.1016/j.desal.2017.04.031

[23] Cambridge ML, Zavala-Perez A, Cawthray GR, Mondon J, Kendrick GA. Effects of high salinity from desalination brine on growth, photosynthesis, water relations and osmolyte concentrations of seagrass *Posidonia australis*. Marine Pollution Bulletin. 2017;**115**:252-260. DOI: 10.1016/j.marpolbul.2016.11.066

[24] Missimer TM, Maliva RG. Environmental issues in seawater reverse osmosis desalination: Intakes and outfalls. Desalination. 2018;**434**:198-215. DOI: 10.1016/j.desal.2017.07.012

[25] Brika B, Omran AA, Dia Addien O. Chemical elements of brine discharge from operational Tajoura reverse osmosis desalination plant. Desalination and Water Treatment. 2016;**57**:5345-5349. DOI: 10.1080/19443994.2014.1003330

[26] Cooley H, Ajami N. Key issues for seawater desalination in California. In: The World's Water. Washington, DC: Island Press/Center for Resource Economics; 2014. pp. 93-121. DOI: 10.5822/978-1-61091-483-3_6

[27] Einav R, Harussi K, Perry D. The footprint of the desalination processes on the environment. Desalination. 2003;**152**:141-154. DOI: 10.1016/S0011-9164(02)01057-3

[28] Belkin N, Rahav E, Elifantz H, Kress N, Berman-Frank I. The effect of

coagulants and antiscalants discharged with seawater desalination brines on coastal microbial communities: A laboratory and in situ study from the Southeastern Mediterranean. Water Research. 2017;**110**:321-331. DOI: 10.1016/j.watres.2016.12.013

[29] Jenkins S, Paduan J, Roberts P, Weis J. Management of Brine Discharges to Coastal Waters: Recommendations of a Science Advisory Panel. Mesa, CA, USA: Southern California Coastal Water Research Project Costa; 2012

[30] Petersen KL, Paytan A, Rahav E, Levy O, Silverman J, Barzel O, et al. Impact of brine and antiscalants on reef-building corals in the Gulf of Aqaba—Potential effects from desalination plants. Water Research. 2018;**144**:183-191. DOI: 10.1016/j.watres.2018.07.009

[31] Water S. Summary of the Environmental Assessment for Public Comment: Sydney's Desalination Project. Sydney, Australia. 2005

[32] Del-Pilar-Ruso Y, Martinez-Garcia E, Giménez-Casalduero F, Loya-Fernández A, Ferrero-Vicente LM, Marco-Méndez C, et al. Benthic community recovery from brine impact after the implementation of mitigation measures. Water Research. 2015;**70**:325-336. DOI: 10.1016/j.watres.2014.11.036

[33] Portillo E, Louzara G, Ruiz de la Rosa M, Quesada J, Gonzalez JC, Roque F, et al. Venturi diffusers as enhancing devices for the dilution process in desalination plant brine discharges. Desalination and Water Treatment. 2013;**51**:525-542. DOI: 10.1080/19443994.2012.694218

[34] Uddin S. Environmental impacts of desalination activities in the Persian Gulf. International Journal of Environmental Science and Development. 2014;**5**:114-117. DOI: 10.7763/IJESD.2014.V5.461

[35] Li H, Shi A, Li M, Zhang X. Effect of pH, temperature, dissolved oxygen, and flow rate of overlying water on heavy metals release from storm sewer sediments. Journal of Chemistry. 2013;**2013**:1-11. DOI: 10.1155/2013/434012

[36] Alshahri F. Heavy metal contamination in sand and sediments near to disposal site of reject brine from desalination plant, Arabian Gulf: Assessment of environmental pollution. Environmental Science and Pollution Research. 2017;**24**:1821-1831. DOI: 10.1007/s11356-016-7961-x

[37] Alharbi T, Alfaifi H, Almadani SA, El-Sorogy A. Spatial distribution and metal contamination in the coastal sediments of Al-Khafji area, Arabian Gulf, Saudi Arabia. Environmental Monitoring and Assessment. 2017;**189**:634. DOI: 10.1007/s10661-017-6352-1

[38] Mohamed AMO, Maraqa M, Al Handhaly J. Impact of land disposal of reject brine from desalination plants on soil and groundwater. Desalination. 2005;**182**:411-433. DOI: 10.1016/j.desal.2005.02.035

[39] Bhandary H, Sabarathinam C, Al-Khalid A. Occurrence of hypersaline groundwater along the coastal aquifers of Kuwait. Desalination. 2018;**436**:15-27. DOI: 10.1016/j.desal.2018.02.004

[40] Pramanik BK, Shu L, Jegatheesan V. A review of the management and treatment of brine solutions. Environmental Science: Water Research & Technology. 2017;**3**:625-658. DOI: 10.1039/C6EW00339G

[41] Ahmed M, Shayya WH, Hoey D, Al-Handaly J. Brine disposal from reverse osmosis desalination plants in Oman and the United Arab Emirates. Desalination. 2001;**133**:135-147. DOI: 10.1016/S0011-9164(01)80004-7

[42] Younos T. Environmental issues of desalination. Journal of Contemporary Water Research & Education. 2009;**132**:11-18. DOI: 10.1111/j.1936-704X.2005.mp132001003.x

[43] Ziolkowska JR. Is desalination affordable?—Regional cost and price analysis. Water Resources Management. 2015;**29**:1385-1397. DOI: 10.1007/s11269-014-0901-y

[44] Shrivastava I, Adams EE. Pre-dilution of desalination reject brine: Impact on outfall dilution in different water depths. Journal of Hydro-Environment Research. 2019;**24**:28-35. DOI: 10.1016/j.jher.2018.09.001

[45] Meneses M, Pasqualino JC, Céspedes-Sánchez R, Castells F. Alternatives for reducing the environmental impact of the main residue from a desalination plant. Journal of Industrial Ecology. 2010;**14**:512-527. DOI: 10.1111/j.1530-9290.2010.00225.x

[46] Lattemann S, Höpner T. Environmental impact and impact assessment of seawater desalination. Desalination. 2008;**220**:1-15. DOI: 10.1016/j.desal.2007.03.009

[47] Chang J-S. Understanding the role of ecological indicator use in assessing the effects of desalination plants. Desalination. 2015;**365**:416-433. DOI: 10.1016/j.desal.2015.03.013

[48] Valipour A, Hamnabard N, Woo K-S, Ahn Y-H. Performance of high-rate constructed phytoremediation process with attached growth for domestic wastewater treatment: Effect of high TDS and Cu. Journal of Environmental Management. 2014;**145**:1-8. DOI: 10.1016/j.jenvman.2014.06.009

[49] Hobbs CM, Arevalo JM, Kiefer CA. Concentrate management: Weigh benefits and drawbacks to ensure success. Opflow. 2016;**42**:22-25. DOI: 10.5991/OPF.2016.42.0054

[50] Ziolkowska JR, Reyes R. Prospects for desalination in the United States—Experiences from California, Florida, and Texas. In: Competition for Water Resources. Netherlands: Elsevier; 2017. pp. 298-316. DOI: 10.1016/B978-0-12-803237-4.00017-3

[51] Saripalli K, Sharma M, Bryant S. Modeling injection well performance during deep-well injection of liquid wastes. Journal of Hydrology. 2000;**227**:41-55. DOI: 10.1016/S0022-1694(99)00164-X

[52] Ahmed M, Shayya WH, Hoey D, Al-Handaly J. Brine disposal from inland desalination plants. Water International. 2002;**27**:194-201. DOI: 10.1080/02508060208686992

[53] Wongpan P, Hughes KG, Langhorne PJ, Smith IJ. Brine convection, temperature fluctuations, and permeability in winter Antarctic land-fast sea ice. Journal of Geophysical Research, Oceans. 2018;**123**:216-230. DOI: 10.1002/2017JC012999

[54] Malakar A, Snow DD, Ray C. Irrigation water quality—A contemporary perspective. Water. 2019;**11**:1482. DOI: 10.3390/w11071482

[55] Rodríguez FA, Santiago DE, Franquiz Suárez N, Ortega Méndez JA, Veza JM. Comparison of evaporation rates for seawater and brine from reverse osmosis in traditional salt works: Empirical correlations. Water Supply. 2012;**12**:234-240. DOI: 10.2166/ws.2012.133

[56] Roychoudhury AN, Petersen J. Geochemical evaluation of soils and groundwater affected by infiltrating effluent from evaporation ponds of a heavy mineral processing facility, West Coast, South Africa.

Journal of Geochemical Exploration. 2014;**144**:478-491. DOI: 10.1016/j.gexplo.2014.02.016

[57] Xu P, Cath T, Wang G, Drewes JE, Ruetten J. Critical Assessment of Implementing Desalination Technology. Virginia, USA. 2009

[58] Mohammadesmaeili F, Badr MK, Abbaszadegan M, Fox P. Mineral recovery from inland reverse osmosis concentrate using isothermal evaporation. Water Research. 2010;**44**:6021-6030. DOI: 10.1016/j.watres.2010.07.070

[59] Mohammadesmaeili F, Badr MK, Abbaszadegan M, Fox P. Byproduct recovery from reclaimed water reverse osmosis concentrate using lime and soda-ash treatment. Water Environment Research. 2010;**82**:342-350. DOI: 10.2175/106143009X12487095236919

[60] Seigworth A, Ludlum R, Reahl E. Case study: Integrating membrane processes with evaporation to achieve economical zero liquid discharge at the Doswell combined cycle facility. Desalination. 1995;**102**:81-86. DOI: 10.1016/0011-9164(95)00044-3

[61] Ahmed M, Arakel A, Hoey D, Thumarukudy MR, Goosen MFA, Al-Haddabi M, et al. Feasibility of salt production from inland RO desalination plant reject brine: A case study. Desalination. 2003;**158**:109-117. DOI: 10.1016/S0011-9164(03)00441-7

[62] Ahmad M, Williams P. Assessment of desalination technologies for high saline brine applications—Discussion paper. Desalination and Water Treatment. 2011;**30**:22-36. DOI: 10.5004/dwt.2011.1374

[63] Kolluri BS. Silica Reduction via Lime Treatment of Brine Concentrates. Texas, USA: The University of Texas at El Paso; 2003

[64] Tarquin AJ. Volume reduction of high silica RO concentrate. In: Membrane Treatment for Drinking Water and Reuse Applications: A Compendium of Peer-Reviewed Papers. Texas, USA. 2006

[65] Gabelich CJ, Williams MD, Rahardianto A, Franklin JC, Cohen Y. High-recovery reverse osmosis desalination using intermediate chemical demineralization. Journal of Membrane Science. 2007;**301**:131-141. DOI: 10.1016/j.memsci.2007.06.007

[66] Duan J, Gregory J. Coagulation by hydrolysing metal salts. Advances in Colloid and Interface Science. 2003;**100-102**:475-502. DOI: 10.1016/S0001-8686(02)00067-2

[67] Stichel W. Water treatment processes; simple options. In: Vigneswaran VS, Visvanathan C, editors. Materials and Corrosion. Vol. 48. USA: CRC Press/Lewis Publishers; 1995/1997. pp. 332-333. DOI: 10.1002/maco.19970480512

[68] Kabsch-Korbutowicz M. Effect of Al coagulant type on natural organic matter removal efficiency in coagulation/ultrafiltration process. Desalination. 2005;**185**:327-333. DOI: 10.1016/j.desal.2005.02.083

[69] Umar M, Roddick F, Fan L. Comparison of coagulation efficiency of aluminium and ferric-based coagulants as pre-treatment for UVC/H_2O_2 treatment of wastewater RO concentrate. Chemical Engineering Journal. 2016;**284**:841-849. DOI: 10.1016/j.cej.2015.08.109

[70] Dialynas E, Mantzavinos D, Diamadopoulos E. Advanced treatment of the reverse osmosis concentrate produced during reclamation of municipal wastewater. Water Research. 2008;**42**:4603-4608. DOI: 10.1016/j.watres.2008.08.008

[71] Comstock SEH, Boyer TH, Graf KC. Treatment of nanofiltration and reverse osmosis concentrates: Comparison of precipitative softening, coagulation, and anion exchange. Water Research. 2011;**45**:4855-4865. DOI: 10.1016/j.watres.2011.06.035

[72] Kazner C, Jamil S, Phuntsho S, Shon HK, Wintgens T, Vigneswaran S. Forward osmosis for the treatment of reverse osmosis concentrate from water reclamation: Process performance and fouling control. Water Science and Technology. 2014;**69**:2431-2437. DOI: 10.2166/wst.2014.138

[73] Bagastyo AY, Keller J, Poussade Y, Batstone DJ. Characterisation and removal of recalcitrants in reverse osmosis concentrates from water reclamation plants. Water Research. 2011;**45**:2415-2427. DOI: 10.1016/j.watres.2011.01.024

[74] Shon HK, Vigneswaran S, Snyder SA. Effluent organic matter (EfOM) in wastewater: Constituents, effects, and treatment. Critical Reviews in Environmental Science and Technology. 2006;**36**:327-374. DOI: 10.1080/10643380600580011

[75] Baudequin C, Couallier E, Rakib M, Deguerry I, Severac R, Pabon M. Purification of firefighting water containing a fluorinated surfactant by reverse osmosis coupled to electrocoagulation-filtration. Separation and Purification Technology. 2011;**76**:275-282. DOI: 10.1016/j.seppur.2010.10.016

[76] Den W, Wang C-J. Removal of silica from brackish water by electrocoagulation pretreatment to prevent fouling of reverse osmosis membranes. Separation and Purification Technology. 2008;**59**:318-325. DOI: 10.1016/j.seppur.2007.07.025

[77] Zhou T, Lim T-T, Chin S-S, Fane AG. Treatment of organics in reverse osmosis concentrate from a municipal wastewater reclamation plant: Feasibility test of advanced oxidation processes with/without pretreatment. Chemical Engineering Journal. 2011;**166**:932-939. DOI: 10.1016/j.cej.2010.11.078

[78] Van Geluwe S, Vinckier C, Braeken L, Van der Bruggen B. Ozone oxidation of nanofiltration concentrates alleviates membrane fouling in drinking water industry. Journal of Membrane Science. 2011;**378**:128-137. DOI: 10.1016/j.memsci.2011.04.059

[79] Broséus R, Vincent S, Aboulfadl K, Daneshvar A, Sauvé S, Barbeau B, et al. Ozone oxidation of pharmaceuticals, endocrine disruptors and pesticides during drinking water treatment. Water Research. 2009;**43**:4707-4717. DOI: 10.1016/j.watres.2009.07.031

[80] These A, Reemtsma T. Structure-dependent reactivity of low molecular weight fulvic acid molecules during ozonation. Environmental Science & Technology. 2005;**39**:8382-8387. DOI: 10.1021/es050941h

[81] Lee LY, Ng HY, Ong SL, Hu JY, Tao G, Kekre K, et al. Ozone-biological activated carbon as a pretreatment process for reverse osmosis brine treatment and recovery. Water Research. 2009;**43**:3948-3955. DOI: 10.1016/j.watres.2009.06.016

[82] Zhang Y, Ghyselbrecht K, Meesschaert B, Pinoy L, Van der Bruggen B. Electrodialysis on RO concentrate to improve water recovery in wastewater reclamation. Journal of Membrane Science. 2011;**378**:101-110. DOI: 10.1016/j.memsci.2010.10.036

[83] Jansen RHS. Ozonation of Humic Substances in a Membrane Contactor Mass Transfer, Product Characterization and Biodegradability. Netherlands: University of Twente; 2005

[84] TAMBO N. Treatability evaluation of general organic matter. Matrix conception and its application for a regional water and waste water system. Water Research. 1978;**12**:931-950. DOI: 10.1016/0043-1354(78)90077-5

[85] Thomson J, Parkinson A, Roddick FA. Depolymerization of chromophoric natural organic matter. Environmental Science & Technology. 2004;**38**:3360-3369. DOI: 10.1021/es049604j

[86] Westerhoff P, Moon H, Minakata D, Crittenden J. Oxidation of organics in retentates from reverse osmosis wastewater reuse facilities. Water Research. 2009;**43**:3992-3998. DOI: 10.1016/j.watres.2009.04.010

[87] Liu K, Roddick FA, Fan L. Potential of UV/H_2O_2 oxidation for enhancing the biodegradability of municipal reverse osmosis concentrates. Water Science and Technology. 2011;**63**:2605-2611. DOI: 10.2166/wst.2011.465

[88] Zhou M, Tan Q, Wang Q, Jiao Y, Oturan N, Oturan MA. Degradation of organics in reverse osmosis concentrate by electro-Fenton process. Journal of Hazardous Materials. 2012;**215-216**:287-293. DOI: 10.1016/j.jhazmat.2012.02.070

[89] Vallero MV, Hulshoff Pol L, Lettinga G, Lens PN. Effect of NaCl on thermophilic (55°C) methanol degradation in sulfate reducing granular sludge reactors. Water Research. 2003;**37**:2269-2280. DOI: 10.1016/S0043-1354(03)00024-1

[90] Häyrynen K, Langwaldt J, Pongrácz E, Väisänen V, Mänttäri M, Keiski RL. Separation of nutrients from mine water by reverse osmosis for subsequent biological treatment. Minerals Engineering. 2008;**21**:2-9. DOI: 10.1016/j.mineng.2007.06.003

[91] Ersever I, Ravindran V, Pirbazari M. Biological denitrification of reverse osmosis brine concentrates: II. Fluidized bed adsorber reactor studies. Journal of Environmental Engineering and Science. 2007;**6**:519-532. DOI: 10.1139/S07-009

[92] Ng HY, Lee LY, Ong SL, Tao G, Viawanath B, Kekre K, et al. Treatment of RO brine–towards sustainable water reclamation practice. Water Science and Technology. 2008;**58**:931-936. DOI: 10.2166/wst.2008.713

[93] Shi X, Lefebvre O, Ng KK, Ng HY. Sequential anaerobic–aerobic treatment of pharmaceutical wastewater with high salinity. Bioresource Technology. 2014;**153**:79-86. DOI: 10.1016/j.biortech.2013.11.045

[94] Lu J, Fan L, Roddick FA. Potential of BAC combined with UVC/H_2O_2 for reducing organic matter from highly saline reverse osmosis concentrate produced from municipal wastewater reclamation. Chemosphere. 2013;**93**:683-688. DOI: 10.1016/j.chemosphere.2013.06.008

[95] Subramani A, Jacangelo JG. Treatment technologies for reverse osmosis concentrate volume minimization: A review. Separation and Purification Technology. 2014;**122**:472-489. DOI: 10.1016/j.seppur.2013.12.004

[96] Cath TY, Gormly S, Beaudry EG, Flynn MT, Adams VD, Childress AE. Membrane contactor processes for wastewater reclamation in space. Journal of Membrane Science. 2005;**257**:85-98. DOI: 10.1016/j.memsci.2004.08.039

[97] McCutcheon JR, McGinnis RL, Elimelech M. A novel ammonia-carbon dioxide forward (direct) osmosis desalination process. Desalination. 2005;**174**:1-11. DOI: 10.1016/j.desal.2004.11.002

[98] McCutcheon JR, McGinnis RL, Elimelech M. Desalination by ammonia-carbon dioxide forward osmosis: Influence of draw and feed solution concentrations on process performance. Journal of Membrane Science. 2006;**278**:114-123. DOI: 10.1016/j.memsci.2005.10.048

[99] Bross L, Krause S, Wannewitz M, Stock E, Sandholz S, Wienand I. Insecure security: Emergency water supply and minimum standards in countries with a high supply reliability. Water. 2019;**11**:732. DOI: 10.3390/w11040732

[100] Tang W, Ng HY. Concentration of brine by forward osmosis: Performance and influence of membrane structure. Desalination. 2008;**224**:143-153. DOI: 10.1016/j.desal.2007.04.085

[101] Holloway R, Childress A, Dennett K, Cath T. Forward osmosis for concentration of anaerobic digester centrate. Water Research. 2007;**41**:4005-4014. DOI: 10.1016/j.watres.2007.05.054

[102] McGinnis RL, Hancock NT, Nowosielski-Slepowron MS, McGurgan GD. Pilot demonstration of the NH_3/CO_2 forward osmosis desalination process on high salinity brines. Desalination. 2013;**312**:67-74. DOI: 10.1016/j.desal.2012.11.032

[103] Hancock NT. High recovery brine treatment using forward osmosis. In: Proceedings of the Membrane Technology Conference. Phoenix, Arizona: American Membrane Technology Association (AMTA)/ American Water Works Association; 2013

[104] Achilli A, Cath TY, Marchand EA, Childress AE. The forward osmosis membrane bioreactor: A low fouling alternative to MBR processes. Desalination. 2009;**239**:10-21. DOI: 10.1016/j.desal.2008.02.022

[105] Mi B, Elimelech M. Organic fouling of forward osmosis membranes: Fouling reversibility and cleaning without chemical reagents. Journal of Membrane Science. 2010;**348**:337-345. DOI: 10.1016/j.memsci.2009.11.021

[106] Lee S, Boo C, Elimelech M, Hong S. Comparison of fouling behavior in forward osmosis (FO) and reverse osmosis (RO). Journal of Membrane Science. 2010;**365**:34-39. DOI: 10.1016/j.memsci.2010.08.036

[107] Drioli E, Di Profio G, Curcio E. Progress in membrane crystallization. Current Opinion in Chemical Engineering. 2012;**1**:178-182. DOI: 10.1016/j.coche.2012.03.005

[108] Di Profio G, Salehi SM, Curcio E, Drioli E. 3.11 Membrane crystallization technology. In: Comprehensive Membrane Science and Engineering. Amsterdam, Netherlands: Elsevier; 2017. pp. 297-317. DOI: 10.1016/B978-0-12-409547-2.12247-4

[109] Ali A, Quist-Jensen C, Macedonio F, Drioli E. Application of membrane crystallization for minerals' recovery from produced water. Membranes (Basel). 2015;**5**:772-792. DOI: 10.3390/membranes5040772

[110] Quist-Jensen CA, Macedonio F, Horbez D, Drioli E. Reclamation of sodium sulfate from industrial wastewater by using membrane distillation and membrane crystallization. Desalination. 2017;**401**:112-119. DOI: 10.1016/j.desal.2016.05.007

[111] Lokare OR, Tavakkoli S, Khanna V, Vidic RD. Importance of feed recirculation for the overall energy consumption in membrane distillation systems. Desalination. 2018;**428**:250-254. DOI: 10.1016/j.desal.2017.11.037

[112] Macedonio F, Drioli E, Gusev AA, Bardow A, Semiat R, Kurihara M. Efficient technologies for worldwide clean water supply. Chemical

Engineering and Processing Process Intensification. 2012;**51**:2-17. DOI: 10.1016/j.cep.2011.09.011

[113] Martinetti CR, Childress AE, Cath TY. High recovery of concentrated RO brines using forward osmosis and membrane distillation. Journal of Membrane Science. 2009;**331**:31-39. DOI: 10.1016/j.memsci.2009.01.003

[114] Camacho L, Dumée L, Zhang J, Li J, Duke M, Gomez J, et al. Advances in membrane distillation for water desalination and purification applications. Water. 2013;**5**:94-196. DOI: 10.3390/w5010094

[115] Song L, Li B, Sirkar KK, Gilron JL. Direct contact membrane distillation-based desalination: Novel membranes, devices, larger-scale studies, and a model. Industrial and Engineering Chemistry Research. 2007;**46**:2307-2323. DOI: 10.1021/ie0609968

[116] Liu K, Roddick FA, Fan L. Impact of salinity and pH on the UVC/ H_2O_2 treatment of reverse osmosis concentrate produced from municipal wastewater reclamation. Water Research. 2012;**46**:3229-3239. DOI: 10.1016/j.watres.2012.03.024

[117] Tun CM, Groth AM. Sustainable integrated membrane contactor process for water reclamation, sodium sulfate salt and energy recovery from industrial effluent. Desalination. 2011;**283**:187-192. DOI: 10.1016/j.desal.2011.03.054

[118] Janson A, Adham S, Benyahia F, Dores R, Husain A, Minier-Matar J. Membrane distillation of high salinity brines using low grade waste heat. In: Proceedings of the Membrane Technology Conference. Phoenix, Arizona: American Membrane Technology Association (AMTA)/ American Water Works Association; 2013

[119] Susanto H. Towards practical implementations of membrane distillation. Chemical Engineering and Processing Process Intensification. 2011;**50**:139-150. DOI: 10.1016/j.cep.2010.12.008

[120] Criscuoli A, Drioli E. Energetic and exergetic analysis of an integrated membrane desalination system. Desalination. 1999;**124**:243-249. DOI: 10.1016/S0011-9164(99)00109-5

[121] Edwie F, Chung T-S. Development of hollow fiber membranes for water and salt recovery from highly concentrated brine via direct contact membrane distillation and crystallization. Journal of Membrane Science. 2012;**421-422**:111-123. DOI: 10.1016/j.memsci.2012.07.001

[122] Dumée LF, Sears K, Schütz J, Finn N, Huynh C, Hawkins S, et al. Characterization and evaluation of carbon nanotube Bucky-paper membranes for direct contact membrane distillation. Journal of Membrane Science. 2010;**351**:36-43. DOI: 10.1016/j.memsci.2010.01.025

[123] Melián-Martel N, Sadhwani JJ, Ovidio Pérez Báez S. Saline waste disposal reuse for desalination plants for the chlor-alkali industry. Desalination. 2011;**281**:35-41. DOI: 10.1016/j.desal.2011.07.040

[124] Tanaka Y, Ehara R, Itoi S, Goto T. Ion-exchange membrane electrodialytic salt production using brine discharged from a reverse osmosis seawater desalination plant. Journal of Membrane Science. 2003;**222**:71-86. DOI: 10.1016/S0376-7388(03)00217-5

[125] Morillo J, Usero J, Rosado D, El Bakouri H, Riaza A, Bernaola F-J. Comparative study of brine management technologies for desalination plants. Desalination. 2014;**336**:32-49. DOI: 10.1016/j.desal.2013.12.038

[126] Badruzzaman M, Oppenheimer J, Adham S, Kumar M. Innovative beneficial reuse of reverse osmosis concentrate using bipolar membrane electrodialysis and electrochlorination processes. Journal of Membrane Science. 2009;**326**:392-399. DOI: 10.1016/j.memsci.2008.10.018

[127] Le Dirach J, Nisan S, Poletiko C. Extraction of strategic materials from the concentrated brine rejected by integrated nuclear desalination systems. Desalination. 2005;**182**:449-460. DOI: 10.1016/j.desal.2005.02.037

[128] Peterskovà M, Valderrama C, Gibert O, Cortina JL. Extraction of valuable metal ions (Cs, Rb, Li, U) from reverse osmosis concentrate using selective sorbents. Desalination. 2012;**286**:316-323. DOI: 10.1016/j.desal.2011.11.042

Chapter 2

Water Treatment and Desalination

Mona M. Amin Abdel-Fatah and Ghada Ahmed Al Bazedi

Abstract

Water covers a large area of the earth that reaches about three quarters of the surface of this planet, but we cannot say that all of this water is fresh or drinkable; according to many statistics, the percentage of fresh water reaches about 1% of the total water on earth. But with the great need for fresh water, whether for drinking or other purposes such as agriculture, the search for water treatment methods has become much larger. One of the most important of these methods that have been developed is desalination of seawater using desalination plants; therefore, we will address here the most important methods used in desalination and water treatment.

Keywords: desalination, water treatment, pretreatment, reverse osmosis, nanofiltration

1. Introduction

Water treatment and purification are the procedures of getting rid of unfavorable chemicals, natural contaminants, as well as suspended solids from water. The aim is to deliver water for certain applications. Water is disinfected for drinking, but water treatment may also be intended for different purposes, including chemical, medical, pharmacological, or other industrial applications. This method includes the physical methods such as sedimentation, distillation, and filtration; biological methods e.g. slow sand filters or biologically activated carbon; chemical methods such as coagulation, flocculation, and chlorination in addition to the use of electromagnetic methods such as ultraviolet light.

Desalination has developed a management water alternative resource by allowing the use of oceans, the greatest reservoirs in the world. Desalination is implemented in more than 100 countries around the world, including the United Arab Emirates, Oman, Malta, Portugal, Greece, Italy, India, China, Spain, Cyprus, Saudi Arabia, Japan, and Australia. Worldwide, the desalination plant produces over 3.5 billion gallons per day of potable water. Seawater desalination technology, reachable for decades, made remarkable strides in many arid areas of the world such as the Middle East, the Mediterranean, and the Caribbean. The potential of an RO desalination plant has increased from 1990 to 2020 [1–3].

Seawater and brackish water are the two most important sources that include desalination technology. The seawater desalination area dominated the international industry specifically due to the outstanding abundance of saltwater resources. Rising wide variety of initiatives in the utility area are expected to improve seawater desalination market.

Brackish water is expected to obtain importance, with an increase in 29% of contracts inside the first half of 2017. However, the presence of constrained reservoirs of brackish water across the globe will probably result in a constrained market boom in this segment.

The water desalination market place worldwide was estimated as USD 15.0 billion in 2017 and is anticipated to make out a wholesome increase with CAGR 7.8% over the projected period. Increasing water crisis throughout the world is the key aspect to drive the market needs for water desalination over the coming years [4, 5].

2. Water treatment technologies

The targets of a water treatment are to eliminate undesirable constituents in the water, making it useful for drinking or matching a specific purpose in industrial applications. Generally, different techniques are presented to eliminate the impurities including microorganisms, suspended solids, and organic materials and other dissolved inorganic or environmental tenacious pharmaceutical contaminants.

The selection of the technique depends on the properties of the water to be treated, the treatment system capital cost, and the requirements predicted of the treated water. The methods discussed below are some of the regularly used methods in water treatment and distillation plants. A few or more may additionally now are not employed according to the treatment plant scale and the characteristics of (feed/source) water [6].

2.1 Pretreatment methods

2.1.1 Pumping

The water to be treated must be drawn out from its source or directed into the piping system or storage tanks by pumps. To evade adjoining contaminants to the water, this bodily substructure ought to be made from excellent materials and assembled so that unplanned contamination does not take place.

2.1.2 Screening

The initial stage in treating surface water is to take away large solids, for example, sticks, debris, leaves, and other massive specks that may affect the following purification/treatment steps. The majority of deep groundwater does not require screening before other purification steps.

2.1.3 pH adjustment

The pH value of natural water is close to 7, while seawater pH values vary from 7.5 to 8.4 in the moderate alkaline range. Water can have extensively fluctuating pH values reliant on the geology of the water aquifer or basin and they have an impact on of contaminant involvements. If the water is acidic (with a pH value less than 7), sodium hydroxide, lime, or soda ash can be employed to raise the pH at some point during the water treatment processes. The addition of lime increases the calcium ion concentration; as a consequence, the water's hardness level is increased. For acidic waters, "compelled draft degasifiers" may be an advantageous method to increase the pH value, by way of stripping "dissolved carbon dioxide" from the treated water.

Increasing the pH value of water (alkaline water) helps "coagulation and flocculation" processes to operate correctly and additionally supports reducing the risk of dissolving "lead" from pipes or from solder used in pipe fittings. Adequate alkalinity furthermore lessens the iron pipe corrosion in water. In some cases, to reduce pH, acid, such as carbonic acid, hydrochloric acid, or sulfuric acid, may also be added to alkaline water. Alkaline water (above pH 7.0) does no longer essentially indicate that some metals like lead or copper will not be dispersed from the plumbing gadget into pumped water. The probability of dissolving hazardous materials like lead in water is limited remarkably to defend surfaces of metal pipes as a result of the capability of water to precipitate "calcium carbonate." The precipitate of calcium carbonate is a function of different parameters including pH, temperature, the mineral content, calcium concentration, and alkalinity.

2.1.4 Coagulation and flocculation

The first step in most regular water treatment tactics is the adding of special chemicals to facilitate the elimination of suspended constituents found in water. These suspended constituents may originate from inorganic sources such as "clay" and "silt" or organic sources such as "algae," "bacteria," "viruses," and other naturally occurring "organic matter." Inorganic and natural particles make contributions to the turbidity and complexion of water.

Adding coagulants of inorganic nature such as "aluminum sulfate (or alum)" or "iron(III) salts" like "iron(III) chloride" may trigger countless instantaneous chemical and physical interactions. By using inorganic coagulants, the particles are neutralized in seconds at a very low cost. Also, iron and aluminum ions start to form precipitates containing metal hydroxide. The metal hydroxide precipitates conglomerate into larger particles under natural tactics such as Brownian action and throughout instigated mixing, which is denoted as the "flocculation process." Amorphous metallic hydroxides are regarded as "floc." Amorphous aluminum and iron(III) large hydroxide molecules adsorb and tangle suspended particles, leading to an easier removal of particles through consequent methods like filtration and sedimentation [6–11].

Aluminum hydroxides are fashioned inside a pretty limited pH range: 5.5 to about 7.7. Iron (III) hydroxide can shape over a higher pH vary consisting of pH degrees lower than are effective for alum, typically: 5.0 to 8.5.

In literature, there is an interesting debate and a relative confusion over the utilization of the two terms, coagulation and flocculation: where does coagulation stop and flocculation take place? In plants used for water purification, a high energy is usually needed, speedy mix unit method (with a short detention time, usually in seconds) whereby the coagulant chemical compounds are delivered accompanied by basins of flocculation (detention instances vary from 15 to 45 minutes) at which large paddles or different gentle mixing low energy units to embellish the formation of floc. So, coagulation and flocculation methods continue once coagulants containing metal salt are added.

2.1.5 Sedimentation

The effectiveness of a sedimentation system depends on the suspended particles' "settling velocity," the volume/area of the tank, and the flow rate through the tank. Design sedimentation tanks are calculated within an overflow rate range between 0.5 and 1.0 gallons per minute per square foot (1.25–2.5 m per hour). Generally, sedimentation tank efficiency is not a feature of retention time or the tank depth. However, the tank should be enough to not disturb the sludge formed by water currents, and settled particle interactions are boosted. Near the sludge level on the bottom of the tank, the particle concentrations in the settled water amplify, and taking into consideration settling velocities can make bigger difference due to the accumulation of suspended particles. Typical retention time for sedimentation process diverges from 1.5 to 4 hours and tank depth varies from 10 to 15 ft. (3–4.5 m) [11].

2.1.6 Dissolved air flotation (DAF)

Dissolved air flotation is regularly employed to treat water when suspended particles do not settle down easily in the water by sedimentation. Following the coagulation and flocculation processes, the treated water is directed to the DAF system tanks. Diffusers of air on the tank's base are used to create excellent bubbles of air attached to floc causing a suspended floating mass of the required floc. The floating floc blanket is eliminated from the floor and clear water is drawn from the base of the tank of DAF. DAF system is essential for water that contains algae blooms with low turbidity and high coloration [12–15].

2.1.7 Filtration

Rapid sand filters are the most frequent types of filters used. Water flows vertically through the sand bed with "activated carbon" layer or "anthracite coal" above it. The upper layer eliminates natural components, affecting taste and odor. The gap within particles of sand is mostly bigger than the smallest particle suspended in water, making simple filtration not adequate. Majority of particles pass through the layers. However, some are trapped in the gap areas or adhere to the sand particles. Effective filtration depends on the volume of the sand filter. The filtration rate of the filter is the key for its proper operation: if the upper layer of sand filter blocks all the particles, the sand filter would rapidly clog [10, 16–18].

The cleaning mechanism of the sand filter is done where water is directed rapidly upward via filtering the opposite direction ("back flushing or backwashing") to dispose of set in or undesired particles. Preceding this step, compressed air may also be used to blow up the compacted media of filter to resource the "backwashing" process, known as air scouring. The contaminated water may be dumped, alongside the sludge produced in the sedimentation tank; alternatively, it could be recycled and mixed with the feedwater inflowing to the treatment plant, which is not regularly used because it reintroduces an expanded volume of contaminated water with bacteria into the feedwater.

Pressure filters are usually used in water treatment plants. It works on the same concept as rapid gravity filters; the main difference is that the medium of the filter is contained inside a vessel where water is enforced under pressure through the filter medium. Benefits of gravity filters are [10, 12, 16–18] as follows:

- Sieves much smaller particles than other sand filters do.
- Sieves effectively all particles larger than their specified pore sizes.
- Water flows through them quite rapidly.
- They can endure a difference of pressure across them of around 2–5 atmospheres.
- Easily cleaned "back flushed."

2.1.8 Storage

Water may be saved/stored in the backside tanks for intervals between few days and several months allowing a natural process of biological purification to occur. This is particularly vital if the treatment is done through slow filters of sand.

Storage tanks also furnish a safeguard against short periods of drought/water shortage or permit maintenance of water supply system at some point of temporary pollution incidents in the supply system.

2.1.9 Prechlorination

In many vegetations, the influent water is used to be chlorinated in order to decrease the boom of fouling organisms on the pipes and the used tanks. Because of the attainable negative first-class effects, this has largely been discontinued.

2.1.10 Other treatment methods

- **Ion exchange:** In these systems, zeolite-packed columns or ion exchange resin is employed to remove undesired ions. The widely used water softening case eliminates $Ca2^{+}$ and Mg2+ ions with Na^{+} or K^{+} ions. In addition, ion exchange resins are furthermore used to hinder heavy metal such as mercury, lead, and arsenic [19–21].

- **Ultrafiltration membranes:** Polymeric or ceramic membranes with microscopic pores are employed to remove dissolved matter averting the coagulant usage. This type of pressure-driven membrane is also used to remove bacteria, viruses, end toxins, and other pathogens, as well as it removes most water turbidity [22–25].

3. Desalination technologies

There is no single technique of desalination. The earliest methods were based totally on distillation or thermal evaporation of seawater on a large scale. Some initial distillation plants have been employed for the desalination brackish water, but high cost prevented substantial implementation of this method in different regions around the world.

The foremost exclusion was few countries in the Arabic Gulf region where excess or less expensive strength is existing. Started in the 1970s, more plant life has been hooked up the use of membranes. Membrane technologies were employed to desalinate both brackish water and seawater, though they are more typically used to desalinate brackish water since cost relatively increases with the water's salt content.

Various membrane technologies can withdraw microorganisms and many natural contaminants. Also, membrane typically has lower capital costs and requires much lower energy compared to thermal systems. However, thermal desalination systems are distinguished with producing water containing decreased salt content compared to membrane (typically much less than 25 mg/l (ppm) total dissolved solids (TDS) as produced in thermal systems in contrast to around 500 ppm in water produced from membrane systems).

The different types of desalination technology include the following:

- Distillation
    - Multistage flash (MSF) distillation
    - Multiple-effect distillation (MED)
    - Vapor-compression (VC)

- Ion exchange
- Membrane processes
    - o Electrodialysis (ED)
    - o Reverse osmosis
    - o Nanofiltration (NF)
    - o Membrane distillation (MD)
    - o Forward osmosis (FO)
- Others

3.1 Multistage flash distillation

Distillation is the most established desalting process where MSF technology is regularly used for seawater desalination. “Dual-purpose” (electric power and potable water production) applications are performed using thermal desalination processes, as well as thermal processes are used for applications that are not applicable to be performed by RO or electrodialysis reversal (EDR), for example, feedwater with high salinity (greater than 50,000 mg/l TDS) or where the feedwater conditions would negatively affect the performance and the membrane life. Recent developments in the MED and VC technologies have led to lower capital costs and reduced amount of auxiliary power consumed, which make these processes cost-effective to MSF distillation [26–28].

Generally, multi-stage flash distillation (MSF) and reverse osmosis (RO) desalination processes account for about 80% of the world’s desalination water production capacity. In the Middle East (particularly Arabian Gulf countries), MSF units are widely used and represent over 40% of the world’s desalinated water production capacity. In North Africa, MSF and RO desalination plants seize about 40% each of the desalination market (shown in **Figure 1**).

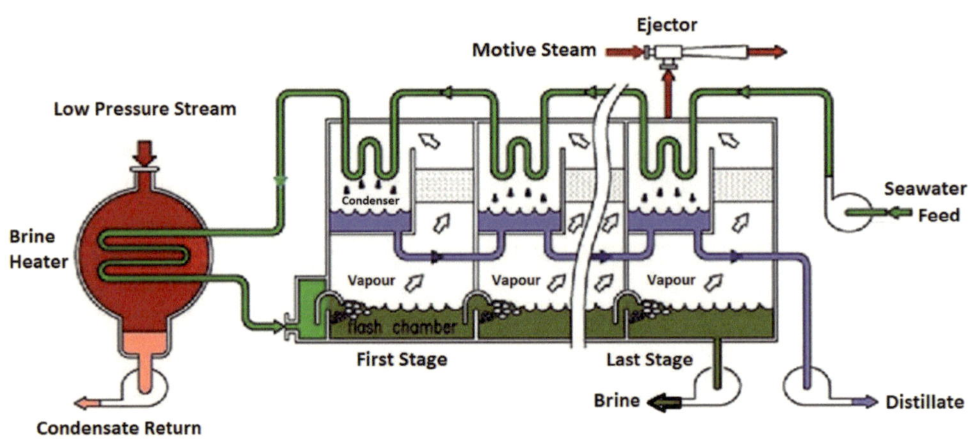

Figure 1.
MSF desalination technology.

3.2 Multiple-effect distillation

The multiple-effect distillation process can be applied in different capacities; for example, small MED plants with a capacity limit to 500 m^3/day were acquainted with the desalination topic in the 1960s. Continuous technological improvements prompt the increase in unit production capacity. The typical number of stages adopted ranges from 8 to 16 stages [28] (shown in **Figure 2**).

3.3 Vapor-compression

Vapor-compression distillation technology is commonly employed for seawater desalting plants within small to medium scale. There are two methods of VC: mechanical vapor-compression (MVC), which is usually electrically driven, or a thermal vapor-compression (TVC). Vapor-compression plants have been built in a variety of configurations [28, 29].

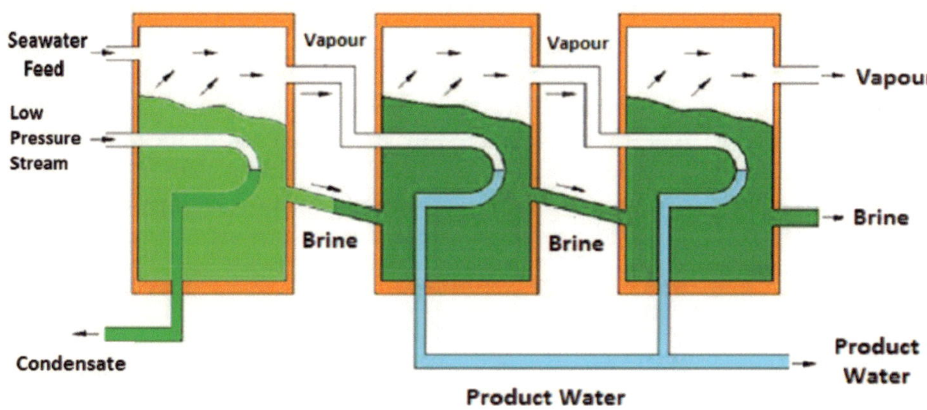

Figure 2.
MED schematic diagram.

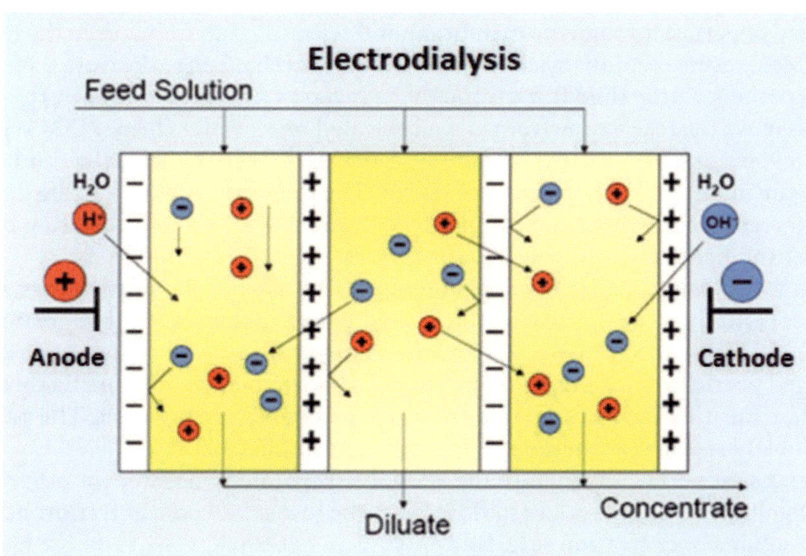

Figure 3.
Ion exchange in electrodialysis unit.

3.4 Electrodialysis and electrodialysis reversal

Electrodialysis desalination process is in some way similar to "ion exchange" treatment process, but it differs in utilizing both cation and anion selective membranes to separate charged ions as shown in **Figure 3**. Improvements to ED process, an electrodialysis reversal process, were introduced where it utilizes regular automatic polarity reversal, which helps in decreasing fouling process [30–32].

Water is handed between a negative electrode and a high-quality electrode. Ion exchange membranes permit solely high-quality ions to transfer toward the negative electrode from the feedwater and negative ions to the positive electrode. Similar to ion exchange treatment process, pure water is produced by deionization. Complete elimination of ions from water is likely to occur if the proper settings are met. Normally, a reverse osmosis system is used for water pretreatment to eliminate natural contaminants, and carbon dioxide is removed with gas switch membranes. Treatment efficiency of 99% is possible if the feed stream is fed to the RO system.

3.5 Reverse osmosis

Reverse osmosis (RO) membrane is known as hyper filtration and is the supreme filtration known. Reverse osmosis allows the removal of small particles and dissolved organic matter. It is also employed to purify different fluids including glycol and ethanol, rejecting other ions and contaminants preventing them from passing through the membrane. Reverse osmosis is commonly used in water treatment. Reverse osmosis is employed to generate water that meets the required specifications needed in place [20, 22, 24].

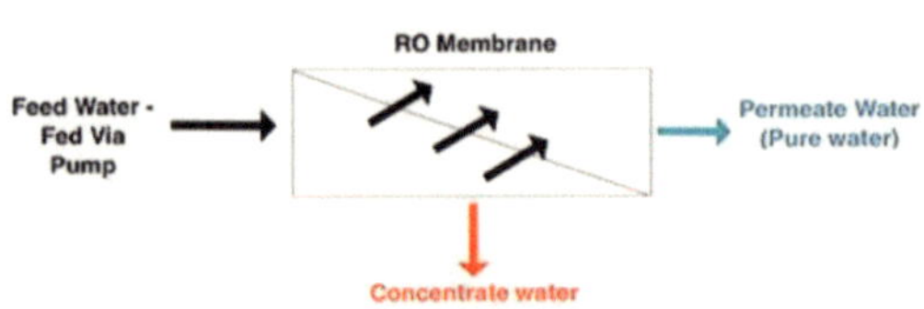

Reverse osmosis membrane is a semipermeable membrane allowing fluid that is to be purified to permit through the membrane and rejecting contaminants in the reject stream. Most reverse osmosis systems use cross flow mechanism to decrease membrane cleaning periods. As the fluid flows through the reverse osmosis membranes, the downstream, remove the reject away from in concentrated reject water (brine) [33–36].

The reverse osmosis process is a pressure-driven process to drive the fluid through the membrane using pressure pump. The pressure increases as the driving forces increase. So, the required driving force increases as the concentration of the reject stream increases.

Reverse osmosis is able to reject proteins, particles, bacteria, salts, sugars, dyes, and other contaminants that are distinguished with a molecular weight greater than the 150–250 daltons range. The reverse osmosis separation of ions is assisted by charged particles. This means that charged dissolved ions will more likely be rejected by the membrane compared to uncharged ones, like organics. The particle will be mostly rejected as the charge and the particle size increases.

When a semipermeable membrane is used to separate two water (or other solvent) volumes, water is going to flow from the low solute concentration side to the high solute concentration side. By applying an external pressure on the higher concentration side, the flow could be stopped or reversed.

In such a case, the phenomenon is called "reverse osmosis." If there are solute molecules only on one side of the system, then the pressure that stops the flow is called the osmotic pressure. The movement of a “solute molecule” within a solvent is overdamped by the solvent molecules that surround it. In fact, the solute movement is wholly determined by fluctuations of the collisions with nearby solvent molecules. So, the average thermal velocity of the molecules is the same as if they were free in the gas phase.

The solute transfers momentum to a wall when the solute is blocked by the wall, and consequently, a pressure on the wall is generated. The pressure on the wall will be the same as the ideal gas pressure of the same molecule concentration, which is attributed to the fact that the velocity is the same as that of free molecule. Therefore, the osmotic pressure π can be calculated by Van’t Hoff formula equation (1):

$$\pi = c\,R\,T \tag{1}$$

where c is the molar solute concentration, R is the gas constant, and T is the absolute temperature. This formula is the same as the pressure formula of the ideal gas.

Due to their modular design concept, RO desalination has a wide range of capacities from large capacities as 395,000 m^3/day to small units down to 0.1 m^3/day.

3.5.1 Problems faced by reverse osmosis membranes

The membrane surface fouling throughout operation reduces the membrane productivity, and if the fouling conditions continue, the salt rejection will suffer. There are three sources for membrane fouling: particles entrained in feedwater, build-up of sparingly soluble minerals, and by-products as a result of growth of microorganisms [25].

A frequent cleaning is required to handle these conditions, which is costly and leads to a shorter service life of the membrane elements. In general, the suspended solids should be eliminated from the feed to the membranes, and for the membrane plant to function well, a suitable pretreatment for the feed is needed. Sparsely soluble minerals are mainly barium and silica, and these contribute to hardness. Microorganism’s growth is most pronounced within the temperature range of 30–45°C [36–38].

3.5.2 Feedwater limitations prior to the RO permeates

The following are the feedwater limitations prior to the RO membranes set by the permeate manufacturers. The pretreatment process should take into consideration the following guidelines that are shown in **Table 1**.

Feed iron, aluminum, and manganese	**Not more than 0.05 mg/l**
Feed bacteriological content	0
Feed chlorine or other oxidants	0
Feed SDI after filtration	<3
Feed organic content (TOC, BOD_5, COD)	0
Feed oil, hydrocarbons, grease content	0
Feed H_2S	0
Suspended solids	<1.0 mg/l
Feed barium, strontium, fluoride	Traces

Table 1.
Feedwater limitations prior to RO permeate.

3.6 Nanofiltration

Nanofiltration is accomplished using a membrane with a pore size of 0.5 and 2 nm and operating pressures between 5 and 40 bars. NF is used to treat solutions that contain organic molecules, sugars, and multivalent salts. Charged nanofiltration membrane rejects ions with negative charge, such as phosphate or sulfate. Non-charged nanofiltration membrane rejects dissolved matter and uncharged in accordance of the shape and size of the molecule, while for positively charged ions, the rejection is affected by the membrane charge and according to the membrane fouling mechanism [17, 32, 34].

Though NF membrane started at the end of the 1970s, NF membrane was used in the 1980s as a separating membrane system, mainly aiming at softening and organic removal. NF membranes have been used commonly in the 1990s. After that, the use ranges of nanofiltration membranes have expanded massively in different applications. Large plants were constructed using NF membranes, for example, the Mery-sur-Oise plant in France (140,000 m^3/day of permeate) in the 1990s.

NF membranes are produced by many membrane manufacturers. The membrane materials are mostly made of polymers like poly-ether-sulfones, aromatic poly(acrylonitrile), polyamides, and poly(phenylene oxide) as well as from different alterations of them. The enhanced permeate flux, the wider membrane pores and less retentive is the membrane.

NF membranes comprise of an active skin layer, which determines the properties of separation, and the structure that supports the porous layer, attributing to the membrane's mechanical strength. The skin layer of the membrane could be connected to the support structure integrally, such as membranes prepared via "phase inversion" (immersion precipitation) process. These membranes have distinguished pores in the nanometer range at the active surface layer. Also, the skin layer can be considered as an extra layer of coating on a tailor-made support structure. The Torell-Meyer-Sievers model (TMC) and the hybrid model assumed a dense skin layer, while the space-charge model and the Donnan-Steric Pore model (DSPM) assumed a porous skin layer. The models were applied in predicting the mechanism of rejection in NF membranes in aqueous solutions [33–35].

NF process is widely used in many industries including treatment of drinking water as well as treatment of wastewater. Operating conditions including temperature, pressure, and pH can affect the energy consumption of NF, which can be reduced by optimizing these parameters. Many researchers studied the effect of operating temperature on the neutral solutes and water transport across NF membranes. The mass transfer and pressure drop are directly affected by the operating parameters and the energy consumption as well. A comparison was done by employing NF process to process streams in the temperature range (T > 50°C) compared to NF at normal temperatures (T 25°C). The high-temperature process stream alters the separation operation attributed to the changes in the active layer of the membrane, which are temperature-dependent. Also, the temperature may affect directly or indirectly a change in viscosity, causing different effects like increased water flux, reduced pressure drop, and increased external mass transfer (lower concentration polarization) [33–35].

3.7 Membrane distillation

Membrane distillation can be defined as a procedure of water desalination membrane presently in constrained commercial use. Membrane distillation represents a hybrid process of distillation and RO using a hydrophobic synthetic membrane to allow the drift of water vapor across the membrane pores and to prevent the solution.

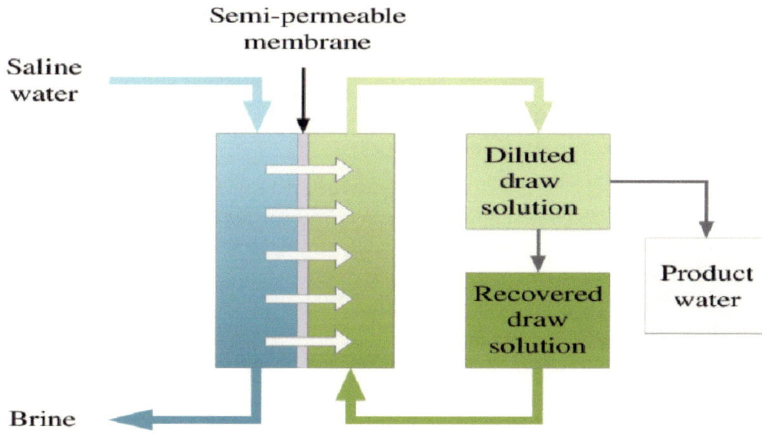

Figure 4.
Forward osmosis.

The vapor pressure difference of the liquid across the membrane is the pressure difference [39–41].

Membrane distillation utilizes the temperature difference across a membrane to vaporize water from a brine solution and condense clear condensate on the chilled side.

3.8 Forward osmosis

Forward osmosis is quite a recent commercial process used for water desalting using a gradient of salt concentration (osmotic pressure) as the driving force through the membrane. If or in case the feed (e.g., seawater) is at one side of the membrane, on the other side of the membrane, there is the higher osmotic pressure "draw" (reusable) solution. As a natural migration process, the water from the feed solution will migrate to the draw solution deprived of employing an external pressure through the membrane. Then, the diluted solution is treated to separate the draw solution from the product [22, 35], as shown in **Figure 4**.

4. Conclusion

We conclude this chapter with a definition of some water treatment and desalination processes used, and we discussed several methods for removing pollutants from water, for example, coagulation and flocculation, sedimentation, clarification, and filtration. As for removing soluble salts, we mentioned some other methods, including ion exchange, thermal processes, and membranes. And also biological issues are removed by disinfection.

Conflict of interest

The authors declare no conflict of interest.

Abbreviations

MSF	multistage flash distillation
MED	multiple-effect distillation

VC	vapor-compression
ED	electro-dialysis
RO	reverse osmosis
NF	nanofiltration
MD	membrane distillation
FO	forward osmosis
EDR	electro-dialysis reversal
TDS	total dissolved solids
MVC	mechanical vapor-compression
TVC	thermal vapor-compression
DAF	dissolved air flotation
π	osmotic pressure
c	molar solute concentration
R	gas constant
T	absolute temperature
TMC	Torell-Meyer-Sievers model
DSPM	Donnan-steric pore model

Author details

Mona M. Amin Abdel-Fatah* and Ghada Ahmed Al Bazedi
Chemical Engineering and Pilot Plant Department, Engineering Research Division, National Research Center (NRC), Dokki, Giza, Egypt

*Address all correspondence to: monaamin46@gmail.com

References

[1] El-Sadek A. Water desalination: An imperative measure for water security in Egypt. Desalination. 2010;**250**(3):876-884. DOI: 10.1016/j.desal.2009.09.143

[2] Growth of desalination and water treatment. Desalination and Water Treatment. 2012;**50**(1-3):1. DOI: 10.1080/19443994.2012.754537

[3] Global Water Desalination Market Size, Analysis, Growth Report, 2014-2025. Available from https://www.hexaresearch.com/research-report/water-desalination-market

[4] Desalination Worldwide. Available from https://www.hbfreshwater.com/desalination-worldwide.html

[5] Seawater and Brackish Water Desalination in the Middle East, North Africa and Central Asia: A Review of Key issues and Experience in Six Countries. 2004. Available from http://siteresources.worldbank.org/INTWSS/Resources/Desal_mainreport-Final2.pdf

[6] Twort AC, Ratnayaka DD, Brandt MJ. Specialized and advanced water treatment processes. Water Supply. 2000:370. https://doi.org/10.1016/B978-034072018-9/50010-2

[7] Altmann J, Zietzschmann F, Geiling E, Ruhl AS, Sperlich A, Jekel M. Impacts of coagulation on the adsorption of organic micro pollutants onto powdered activated carbon in treated domestic wastewater. Chemosphere. 2015;**125**:198-204. DOI: 10.1016/j.chemosphere.2014.12.061

[8] Parsons SA, Jefferson B. Coagulation and flocculation. Introduction to Potable Water Treatment Processes. Blackwell Publishing Ltd; 2009. pp. 26-42. ISBN: 978-1-405-12796-7. DOI: 10.1002/9781444305470.ch3

[9] Katrivesis F, Karela A, Papadakis V, Paraskeva C. Revisiting of coagulation-flocculation processes in the production of potable water. Journal of Water Process Engineering. 2009;**27**:193-204. DOI: 10.1016/j.jwpe.2018.12.007

[10] Guo D, Wang H, Fu P, Huang Y, Liu Y, Lv W, et al. Diatomite precoat filtration for wastewater treatment: Filtration performance and pollution mechanisms. Chemical Engineering Research and Design. 2018;**137**:403-411. DOI: 10.1016/j.cherd.2018.06.036

[11] Liu F, Zhang C, Zhao T, Zu Y, Wu X, Li B, et al. Effects of phosphate on the dispersion stability and coagulation/flocculation/sedimentation removal efficiency of anatase nano-particles. Chemosphere. 2019;**224**:580-587. DOI: 10.1016/j.chemosphere.2019.02.162

[12] Russell D. Dissolved air flotation and techniques. In: Practical Wastewater Treatment. 2019. pp. 409-417. https://doi.org/10.1002/9781119527114.ch19

[13] Kwak D, Lee K. Enhanced phosphorus removal in the DAF process by flotation scum recycling for advanced treatment of municipal wastewater. Water Science and Technology. 2015;**72**(4):600-607. DOI: 10.2166/wst.2015.243

[14] Maeng M, Kim H, Lee K, Dockko S. Effect of DAF configuration on the removal of phosphorus and organic matter by a pilot plant treating combined sewer overflows. International Biodeterioration & Biodegradation. 2017;**124**:17-25. DOI: 10.1016/j.ibiod.2017.07.017

[15] Younker JM, Walsh ME. Novel pre-treatment for dissolved air flotation treatment of produced water. Proceedings of the Water Environment Federation. 2013;**19**:380-385. DOI: 10.2175/193864713813667809

[16] Ncube P, Pidou M, Stephenson T, Jefferson B, Jarvis P. Consequences of pH change on wastewater depth filtration using a multimedia filter. Water Research. 2018;**128**:111-119. DOI: 10.1016/j.watres.2017.10.040

[17] Zahrim A, Hilal N. Treatment of highly concentrated dye solution by coagulation/flocculation–sand filtration and nanofiltration. Water Resources and Industry. 2013;**3**:23-34. DOI: 10.1016/j.wri.2013.06.001

[18] Zahrim AY, Tizaoui C, Hilal N. Removal of highly concentrated industrial grade leather dye: Study on several flocculation and sand filtration parameters. Separation Science and Technology. 2011;**46**(6):883-892. DOI: 10.1080/01496395.2010.550596

[19] De Abreu Domingos R, Da Fonseca FV. Evaluation of adsorbent and ion exchange resins for removal of organic matter from petroleum refinery wastewaters aiming to increase water reuse. Journal of Environmental Management. 2018;**214**:362-369. DOI: 10.1016/j.jenvman.2018.03.022

[20] Gu J, Liu H, Wang S, Zhang M, Liu Y. An innovative anaerobic MBR-reverse osmosis-ion exchange process for energy-efficient reclamation of municipal wastewater to NEWater-like product water. Journal of Cleaner Production. 2019;**230**:1287-1293. DOI: 10.1016/j.jclepro.2019.05.198

[21] Wawrzkiewicz M, Hubicki Z. Anion exchange resins as effective sorbents for removal of acid, reactive, and direct dyes from textile wastewaters. Ion Exchange - Studies and Applications. 2015. DOI: 10.5772/60952

[22] Heo J, Kim S, Her N, Park CM, Yu M, Yoon Y. Removal of contaminants of emerging concern by FO, RO, and UF membranes in water and wastewater. Contaminants of Emerging Concern in Water and Wastewater. Advanced Treatment Proces. 2020:139-176. DOI: 10.1016/b978-0-12-813561-7.00005-5

[23] Horstmeyer N, Thies C, Lippert T, Drewes JE. A hydraulically optimized fluidized bed UF membrane reactor (FB-UF-MR) for direct treatment of raw municipal wastewater to enable water reclamation with integrated energy recovery. Separation and Purification Technology. 2020;**235**:116165. DOI: 10.1016/j.seppur.2019.116165

[24] Petrinic I, Korenak J, Povodnik D, Hélix-Nielsen C. A feasibility study of ultrafiltration/reverse osmosis (UF/RO)-based wastewater treatment and reuse in the metal finishing industry. Journal of Cleaner Production. 2015;**101**:292-300. DOI: 10.1016/j.jclepro.2015.04.022

[25] Sioutopoulos D, Karabelas A, Yiantsios S. Organic fouling of RO membranes: Investigating the correlation of RO and UF fouling resistances for predictive purposes. Desalination. 2010;**261**(3):272-283. DOI: 10.1016/j.desal.2010.06.071

[26] Al-Weshahi MA, Anderson A, Tian G. Exergy efficiency enhancement of MSF desalination by heat recovery from hot distillate water stages. Applied Thermal Engineering. 2013;**53**(2):226-233. DOI: 10.1016/j.applthermaleng.2012.02.013

[27] Al-Weshahi MA, Tian G, Anderson A. Performance enhancement of MSF desalination by recovering stage heat from distillate water using internal heat exchanger. Energy Procedia. 2014;**61**:381-384. DOI: 10.1016/j.egypro.2014.11.1130

[28] Gebel J. Thermal desalination processes. In: Kucera J, editor. Desalination Water from Water. 2019:51-138. ISBN: 978-1-118-20852-6. DOI: 10.1002/9781119407874.ch2

[29] Shen J, Feng G, Xing Z, Wang X. Theoretical study of two-stage

water vapor compression systems. Applied Thermal Engineering. 2019;**147**:972-982. DOI: 10.1016/j.applthermaleng.2018.11.012

[30] Nayar KG, Lienhard JHV. Brackish water desalination for greenhouse agriculture: Comparing the costs of RO, CCRO, EDR, and mono-valent-selective EDR. Desalination. 2020;**475**:114188. DOI: 10.1016/j.desal.2019.114188

[31] Turek M, Was J, Dydo P. Brackish water desalination in RO–single pass EDR system. Desalination and Water Treatment. 2009;**7**(1-3):263-266. DOI: 10.5004/dwt.2009.710

[32] Zhang Y, Liu L, Du J, Fu R, Van der Bruggen B, Zhang Y. Fracsis: Ion fractionation and metathesis by a NF-ED integrated system to improve water recovery. Journal of Membrane Science. 2017;**523**:385-393. DOI: 10.1016/j.memsci.2016.09.052

[33] Nikbakht Fini M, Madsen HT, Muff J. The effect of water matrix, feed concentration and recovery on the rejection of pesticides using NF/RO membranes in water treatment. Separation and Purification Technology. 2019;**215**:521-527. DOI: 10.1016/j.seppur.2019.01.047

[34] Mansell B, Ackman P, Tang C, Friess P, Fu P. Pilot-scale testing of a high recovery NF/RO integrated treatment system for indirect potable reuse. Proceedings of the Water Environment Federation. 2011;**14**:3019-3034. DOI: 10.2175/193864711802721055

[35] Altaee A, Hilal N. High recovery rate NF–FO–RO hybrid system for inland brackish water treatment. Desalination. 2015;**363**:19-25. DOI: 10.1016/j.desal.2014.12.017

[36] Jin X, Jawor A, Kim S, Hoek EM. Effects of feed water temperature on separation performance and organic fouling of brackish water RO membranes. Desalination. 2009;**239**(1-3):346-359. DOI: 10.1016/j.desal.2008.03.026

[37] Ruiz-García A, Melián-Martel N, Mena V. Fouling characterization of RO membranes after 11 years of operation in a brackish water desalination plant. Desalination. 2018;**430**:180-185. DOI: 10.1016/j.desal.2017.12.046

[38] Stoica IM, Vitzilaiou E, Lyng Røder H, Burmølle M, Thaysen D, Knøchel S, et al. Biofouling on RO-membranes used for water recovery in the dairy industry. Journal of Water Process Engineering. 2018;**24**:1-10. DOI: 10.1016/j.jwpe.2018.05.004

[39] Damtie MM, Kim B, Woo YC, Choi J. Membrane distillation for industrial wastewater treatment: Studying the effects of membrane parameters on the wetting performance. Chemosphere. 2018;**206**:793-801. DOI: 10.1016/j.chemosphere.2018.05.070

[40] Duke MC, Dow N. Membrane distillation for industrial water treatment. Membrane Distillation. In: Duke MC, Dow N, editors. Membrane Distillation for Industrial Water Treatment: Experiences from Pilot Trials; 2019:371-396. Chapter 16. DOI: 10.1201/9780429287879-16

[41] Benyahia F. Membrane distillation performance analysis. In: Benyahia F, editor. Membrane-Distillation in Desalination. 2019:73-100. Chapter 5. https://doi.org/10.1201/9781315117553

Chapter 3

Water Demand and Salinity

Ayyam Velmurugan, Palanivel Swarnam, Thangavel Subramani, Babulal Meena and M.J. Kaledhonkar

Abstract

The fresh water constitute only 3% of the total water on earth out of which underground water constitute 29 and <1% is in the form of lakes and rivers on the earth surface. Considering the rapidly increasing human population and demand for diverse food items crop production must increase substantially. At the same time arable land and good quality irrigation water resources are being depleted at faster rate particularly in the arid, semi-arid and tropical regions. Over the years the salinization of soil and water has steadily increased due to various causes and the increase in food production has essentially depends on this degrading resources. Since the balance between water demand and water availability has reached critical level in many regions of the world a sustainable approach to water resources and salinity management has become imperative. This chapter highlights global water resources, its demand and supply, salinity and its causes, effect of climate change and its management for sustainable use.

Keywords: water resources, demand, salinity, aquifers, climate change

1. Introduction

Water is the most important resource essential for sustenance of life on earth and drive the economic development of human society. Nearly 97% of the total water on earth is in oceans in the form of saline water and only the remaining 3% is fresh water. Out of this, nearly 70% of fresh water is in the form of ice present in the polar region and higher mountain ranges. Underground water constitutes 27% of fresh water and only <1% is in the form of surface water present in lakes and rivers. Rapid changes in human lifestyle coupled with urbanization and industrialization has created pressure on the limited fresh water resources. Further, the impending climate change has favoured salinization of both land and water across many parts of the world [1].

The projection given by FAO indicated increase of food requirements as a result of burgeoning population by 20% in developed countries and 60% in developing countries. In other words food requirements are increasing quicker than crop production. Hence, there is urgent need to improve alternative agricultural strategies [2, 3]. Among the many reasons affecting agricultural productivity in the tropical region, salinity and associated factors, like waterlogging and/or drought exaggerated by climate change have contributed significantly. The increase in saline areas has been directly attributed to both water and soil salinity problems. In the coastal areas inundation of low-lying areas by sea water and sea water intrusion into the fresh water aquifers contribute to the coastal salinization.

Since the balance between water demand and water availability has reached critical levels in many regions of the world and increased demand for water and food production is likely in the future, a sustainable approach to water resource use and salinity management has become imperative [4]. A number of approaches have been developed to combat the salinity problems and increase the food grain production, based on specific types of site, regional and global problems. This chapter highlights concepts of water resources, its availability, human demand and use of fresh water, the effect of climate change and other factors on salinity and water resources in the future and also discusses the ways to manage this precious natural resource.

2. Global water resources

The concept of water resources encompasses qualitative socio-economic and environmental dimensions besides its quantitative and physical aspects. The source of all forms of water either directly or indirectly is precipitation, often used interchangeable with total rainfall in literatures. However, with reference to water resources, precipitation is considered as gain, evapotranspiration is viewed as loss and human use including for agriculture is described as demand. When the resources are contaminated by human activities or turned into saline by natural means, the fresh water resources get reduced which intensify the water demand. At the same time part of the rainfall after reaching the ground get evapotranspired or moves to the fresh water resources (surface and ground water).

There are several reports on global total fresh water resources which are estimated with reference to a particular year, and may vary with the progress of time as it is depend on several dynamic components. The total freshwater resources spread across the world are estimated to be in the order of 43,750 km^3 $year^{-1}$. At the continental level on an average America has the largest share of the world's total freshwater resources with 45%, followed by Asia (28%), Europe (16%), and Africa (9%) [5]. Due to uneven distribution of population and water resources, continent wise estimation of water resource per inhabitant showed that America has highest amount with 24,000 m^3 $year^{-1}$ followed by Europe (9300 m^3 $year^{-1}$), Africa (5000 m^3 $year^{-1}$) and Asia (3400 m^3 $year^{-1}$) [6]. At the regional level, tropical humid region has fairly good IPWR per capita due to good amount of annual rainfall and water resources. However at the country level, countries located in the Arabian Gulf and Northern Africa (Morocco, Algeria, Bahrain, Jordan, Kuwait, Libyan Arab Jamahiriya, Maldives, Malta, Qatar, Saudi Arabia, United Arab Emirates and Yemen) are having very low total renewable water resources (TRWR) of 500 m^3 per inhabitant. In terms of internal renewable resources (IRWR), the threshold of 1000 m^3 per inhabitant is considered as water stress, which showed that countries located in North Africa and the Middle East are at the most critical stress level with values ranging from 0 to 1000 m^3 $year^{-1}$ per person.

3. Global water withdrawal

Water is withdrawn from the available resources for various purposes which creates the demand. In other words, to understand the relation between supply and demand, the ratio between water withdrawal by agriculture, municipalities and industries over total renewable water resources is used. This also indicates the level of human pressure on water resources. Arid and semi-arid regions in Asia and Africa have maximum withdrawal of more than 90% of renewable water as given in

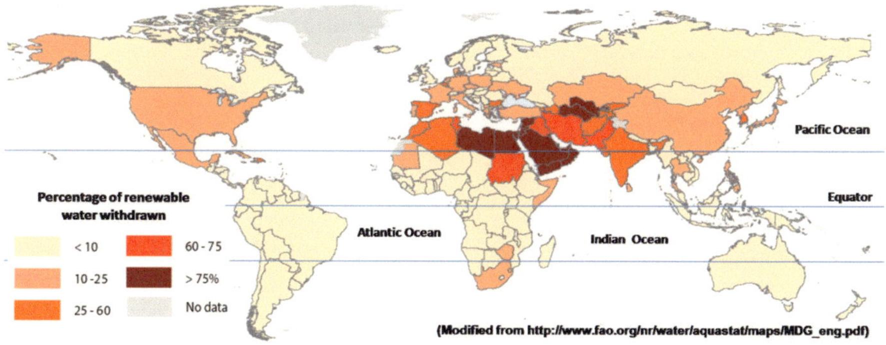

Figure 1.
Percentage of renewable water resources withdrawn.

Figure 1. In these areas surface flow is seasonal due to less rainfall. As a consequence this region exploits more ground water resources than other region.

Agriculture accounts for roughly 70% of total freshwater withdrawals globally and for over 90% in the majority of Least Developed Countries (LDCs) [7]. If remedial measures to improve the efficiency are not seriously implemented, by 2050 global agricultural water consumption is projected to increase by about 20%. Globally, some 38% of irrigated areas depend on groundwater [8] which has contributed to a 10-fold increase of groundwater abstraction for agricultural irrigation over the last 50 years. Conversely, almost half of the world's population depends on groundwater for drinking consequently salinization or overexploitation will affect the freshwater availability for domestic purpose.

4. Characterizing salinity

In the context of global water resources, demand and salinity, it is imperative to define salinity as it various in intensity and the severity is different based on the intended purpose. **Salinity** is a measure of the content of salts in soil or water. Salts are highly soluble in surface and groundwater and can be transported with water movement. There are two kinds of salinity viz., primary and secondary salinity.

- **Primary salinity** is produced by natural processes such as weathering of rocks or wind and rain depositing salt over thousands of years. Nevertheless, the distribution of salt deposits in the world natural region is uneven, and the impacts of salinity vary due to different topography and the age of the landscapes.

- **Secondary salinity** occurs due to the accumulation of salt from the primary source. This may be due to extensive land clearing and unjust land use practices. This is mostly observed in the form of 'dryland salinity' or 'irrigation-induced salinity.' Dryland salinity occurs due to the replacement of deep-rooted native plants by shallow-rooted plants having less water requirement. In addition, farmers apply excess irrigation water once the irrigation water is made available to them. Consequently this leads to raising the water table and bringing salt to the surface where it can be left behind as the water evaporates. On the other hand, irrigation-induced salinity occurs when excess water applied to crops travels past the root zone to groundwater, raising the water table and salt to the surface. Salt may also be transported across surface and groundwater systems.

Water class	Electrical conductivity (dS/m)	Salt concentration (mg/l)	Type of water
Non-saline	<0.7	<500	Drinking and irrigation water
Slightly saline	0.7–2	500–1500	Irrigation water
Moderately saline	2–10	1500–7000	Primary drainage water and groundwater
Highly saline	10–25	7000–15,000	Secondary drainage water and groundwater
Very highly saline	25–45	1 5000–35,000	Very saline groundwater
Brine	>45	>45,000	Seawater

Table 1.
Classification of saline waters.

The term salinity used herein refers to the total dissolved concentration of inorganic ions (Na, Ca, Mg, K, HCO_3, SO_4 and Cl) in irrigation, drainage and ground waters. Individual concentrations of these cations and anions in a unit volume of water can be expressed either on a chemical equivalent basis ($mmol_c/l$) or on a mass basis (mg/l). The total salt concentration in water is expressed either in terms of the sum of either the cations or anions ($mmol_c/l$) or the sum of cations plus anions (mg/l). But for analytical convenience, salinity is measured as electrical conductivity (EC) expressed in units of (dS/m) [6]. As the solubility of salt vary at different temperature, EC values are always expressed at a standard temperature of 25°C to enable comparisons. This also helps to back convert EC into total salt concentration which is 1 dS/m = 10 $mmol_c/l$ = 700 mg/l. In spite of certain shortcomings, EC is a fair indicator of salinity as plants are mainly sensitive to total salt concentration rather than to the proportions of individual salt constituents. At the same time, for comparison purposes, 'soil salinity', is commonly expressed in terms of the electrical conductivity of an extract of a saturated paste (EC_e; in dS/m) made using a sample of the soil. Normally due to the involvement of different modifying factors rigid water quality classifications are not advised but for the purpose of identifying the levels of water salinities water classification scheme is used.

In terms of total salt concentration, which is the major water quality factor generally limiting the use of waters for crop production and other purposes, water classes are defined (**Table 1**). As per this scheme, only very hardy and tolerant crops can be successfully grown with waters having salinity of 10 dS/m in EC or more. Many drainage waters, including shallow ground waters underlying irrigated lands, fall in the range of 2–10 dS/m in EC. Such waters have good potential for selected crop production with suitable salinity management practices. Reuse of second-generation drainage waters for irrigation in selected locations is sometimes possible particularly for purposes of reducing drainage volume in preparation for ultimate disposal or treatment. Such waters will generally have ECs in the range 10–25 dS/m. Very highly saline waters (25–45 dS/m in EC) and brine (>45 dS/m in EC) are beyond the scope of these guidelines and their uses for crop production are therefore not discussed herein. In summary water having EC more than 10 are not recommended for irrigation and water with EC value <10 are used with suitable salinity management methods.

5. Status of saline land and water

The availability of fresh water for farming is an essential condition for achieving satisfactory and profitable yields, both in terms of unit yields and quality. In coastal regions due to excessive withdrawal of ground water, high evapotranspiration, rise of saline ground water and sea water intrusion pose major challenge. The most common reasons for the increase in salt-affected lands are the mismanagement of irrigated areas. Increase in groundwater pumping results in the intrusion of seawater into the fresh water aquifers. In certain region/islands due to the exhaustion of fresh water aquifers the overlying saline water layers mix with fresh water, resulting in the increase of salinity in the groundwater. In the dry region, high rates of irrigation water application and inadequate or absence of drainage systems has resulted in the movement and deposition of salt on the surface of the soil profile favoured by high evapotranspiration rates. As a result nearly 5–10% of the existing fresh water resources are getting salinized. The critical values of renewable fresh water resources and economic water scarcity and salinization indicate the necessity for regional water use policy and appropriate water management strategies at various levels.

As discussed above the salinity level, both soil and water, has been increasing in many of the regions particularly in the tropics and arid regions though the processes of occurrence of salt affected soils are different. Salinity can be found in different altitudes, from territories below sea level, e.g. the district of the Dead Sea, to mountains rising over 5000 m as the Tibetan Plateau of the Rocky Mountains [9]. Older estimates [10] suggest that 10% of the total arable land is affected by salinity and sodicity, extending over more than 100 countries occupying different proportions of their territory. The description of the types of salt-affected soils, causes of formation and hypothetical salinization cycle has been reported by many researchers [11].

Due to the non-availability of updated information or lack of compilation of regional level assessments the current extent of salt-affected soils are unknown. Based on the FAO/UNESCO Soil Map of the World, Massoud [12] made an estimate of 880 M ha of salt-affected soils of which 36% are in developing countries. These are the potential areas where land can be leased for food production or alternate energy sources using suitable technologies which are currently available. However, Balba [13] gave a global estimate of only 600 M ha as salt-affected soils which included 340 M ha in Asia, 140 M ha in Australia, 60 M ha in South America 30 M ha in Africa, 26 M ha in North America and 1 M ha in Europe. The recent estimate quoted 954.8 M ha of salt-affected soils which is much higher than the previous estimates [14].

Similar to that of soils, the availability of freshwater is a major limiting factor for sustainable agriculture and other developmental activities. In certain regions of the world the water crisis is so severe than the availability of land. Unlike soils, there are several assessments and projection for future water requirements and availability. Global assessment of water availability and projections found decrease in water availability in the developing regions with increasing population pressure [15, 16]. The assessment grouped the water scarcity into physical water scarcity, approaching physical water scarcity and economic water scarcity to understand the water shortage which includes all purpose of water (**Figure 2**). Physical water scarcity means water resource development is approaching or has exceeded sustainable limits. Here water availability is related to water demand which implies that dry areas are not necessarily water scarce. This physical scarcity analysis showed that more than 75% of river flows are withdrawn for agriculture, industry and domestic purposes. In approaching physical water scarcity nearly 60% of river flows are withdrawn and these basins may experience physical water scarcity in the future. The situation may get worsen with more withdrawal to produce more food. Whereas in economic water scarcity even though water in nature is available locally to meet human demands factors such as human, institutional and financial capital limits access to water. The tropical developing region mostly faces the challenges of water scarcity. Economic water scarcity is the major limits for production in sub-Saharan Africa while physical water scarcity limits the production in South Asia.

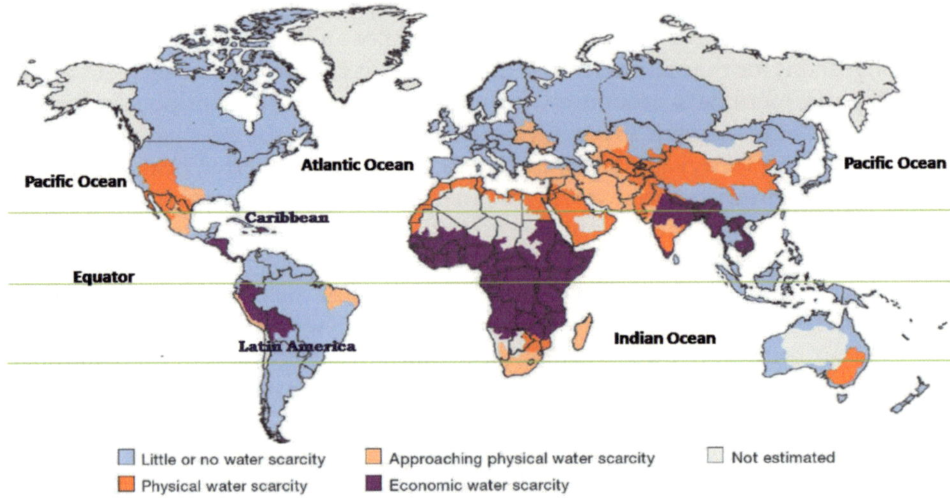

Figure 2.
Areas of physical and economical water scarcity at the basin level in 2007 [16].

In contrast to the crisis, some countries have developed technologies to utilize the saline water. For example, in Israel, farmers carryout crop production with unconventional water resources irrigation and desalination plants have been installed to get fresh water from saline water. In some tropical regions of Asia technologies have been developed to address this issue and we can find agriculture practices based on alternative plant species, most of them are halophytes, which are able to tolerate high temperatures and/or low water availability [17]. Similar attempts are being made in some of the South Asian countries to meet the challenges but in many cases these are at experimental stage.

6. Factors affecting aquifer and salinization

6.1 Land subsidence

Large scale withdrawal of ground water (over exploitation), especially from the artesian aquifers can sometimes result in local land subsidence due to compression of the aquifers. Land subsidence poses serious problems to buildings, other structures and affects the equilibrium of freshwater-sea water interface region. Sometimes this causes inundation of low lying areas, resulting in sea water ingress. The subsidence depends on the nature of sub surface formations, their extent, magnitude and duration of the artesian pressure decline.

6.2 Sea water intrusion

This is one of the most serious emerging problems in the coastal regions. When groundwater is pumped out of coastal aquifers which is in hydraulic connection with the sea due to gradients salt water from the sea may flow towards the well (**Figure 3**). There is a dynamic equilibrium in the seawater-fresh water interface which gets disturbed due to over exploitation of ground water or reduced freshwater recharge. This result in movement of salt water into freshwater aquifers under the influence of groundwater development or by over exploitation which is known

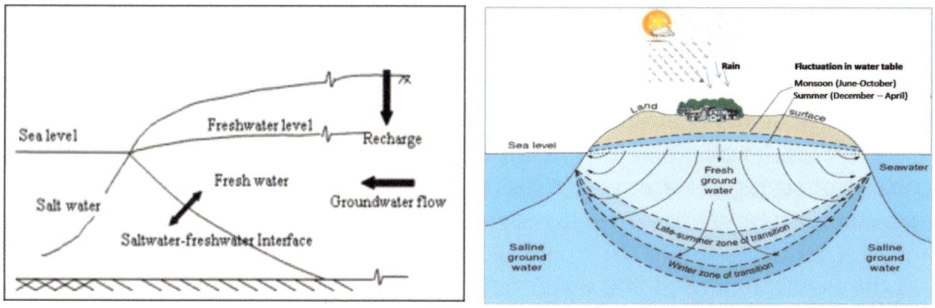

Coastal areas   Island

Figure 3.
Fresh water-salt water interface.

as seawater intrusion. Sometime there is a propensity to point out the occurrence of any saline water along the coastal zone to sea water intrusion. But there may be many reasons for the occurrence of salinity. In order to avoid mistaken diagnoses of seawater intrusion as evidenced by temporary increases of total dissolved salts, chloride-bicarbonate ratio as a criterion to evaluate the intrusion.

6.3 Upcoming of saline water

This phenomenon occurs due to the local rise of the interface between fresh and saline water. This happens when an aquifer having underlying layer of saline water is pumped by a well penetrating only the upper freshwater portion of the aquifer. This rise in interface layer below the well due to excess removal of water is called upcoming of saline water. Generally the interface lies near horizontal at the start of pumping which rises to progressively higher levels with continued pumping of water until eventually it reaches the well. At that point it necessitates closing of the well because of the degrading influence of the saline water. When pumping is stopped, the denser saline water tends to settle downward and to return to its former position. In such areas, the rainwater tend to float over saline water as a thin lens and in such conditions the saline water rises by 40 units for every unit of the fresh water withdrawn. Because of this very fragile ground water system of small islands the fresh water needs to be skimmed to prevent upcoming.

6.4 Geogenic salinity

This kind of salinity is a common water quality problem observed in the coastal aquifers. In these aquifers, the salinity is caused because of leaching of salts in the aquifer material. In certain areas the formation water gets freshened regularly due to the leaching effect. This happens mostly for the water soluble salts only.

6.5 Sea level rise (SLR)

The observed and projected increase in mean sea level due to global warming poses a serious threat to the coastal aquifers particularly located in the small islands. The projected SLR will drive the fresh water-seawater interface more towards inland along coastal aquifers and consequently submerge the lowlying areas with saline sea water. This will result in direct salinization of shallow coastal aquifers. Water resources of the small islands located in the tropical region (Indian Ocean and Polynesian islands) will be significantly affected by the rise in sea level and with the change in rainfall pattern the negative effect will be even greater.

7. Climate change and future water demand

Water has become a scarce natural commodity due to its declining availability and increase in demand for various purposes. This has created huge pressure on the available fresh water resources around the globe. Several reports state that the magnitude of stress on water resources is expected to increase as a consequence of climate change, population growth, economic development and land-use change including urbanization [18]. In consequence several studies were carried out focusing on the assessment of global water demand and its availability. In reality water demand has reached critical levels in several parts of the world, particularly in countries with very limited water availability. Many researchers have concluded that besides climate change, misuse of water, over exploitation and limited infrastructures for water supply are the major reasons for water scarcity.

Globally water consumption for all sectors amounts to 9% of total freshwater resources in the world with agriculture being the largest user, in turn accounting for approximately 70% of total water withdrawals which is equivalent to 2700 km^3 $year^{-1}$ [19]. Agricultural sector receives up to two-thirds of the total water withdrawals and accounts for almost 90% of the total water consumption in the world [5]. As more than 80% of global agricultural land is rainfed water demand is met mostly from the green water resource [16]. In Asia, Africa, Central and South America, the values for specific water withdrawal range from 50 to 100% which experiences great diversity in climatic conditions. Irrigation water withdrawals range between 96 km^3 in Sub-Saharan Africa and 708 km^3 in East Asia; the highest values for specific water withdrawal are observed in South Asia, with 913 km^3 [20].

Analysis of factors affecting water supply and demand indicated that the water demand will be influenced by population growth, industrial development and

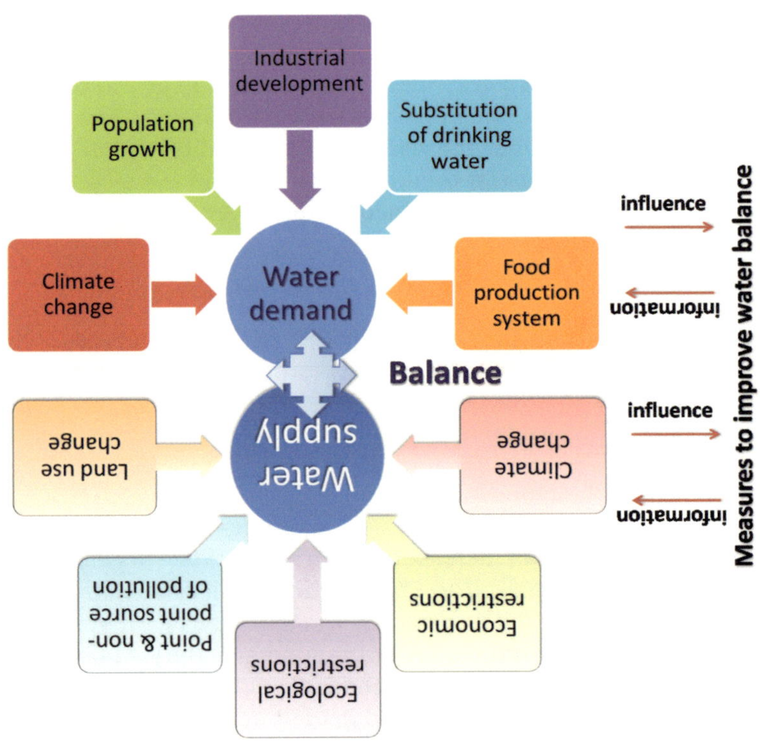

Figure 4.
Driving forces of future water supply and demand (modified from Hornbogen and Schultz [21]).

food production besides climate change. At the same time the water supply will be decided by land use change, ecological and economic restriction, pollution besides climate change (**Figure 4**). The balance between these two will decide the fresh water availability [21]. The future global water situation and development until 2025 was analysed with different scenarios. Under the business as usual scenario the present contrast in water situation between industrialized and developing countries is likely to continue in the future. Withdrawal of water from the available resources in most of the industrialized countries is projected to decline or will remain at the present level due to technological and efficient water management. Consequently the pressure on available freshwater resources will decline. In contrast withdrawal will continue to grow in developing countries due to urbanization and industrialization. Further, the push for development will also be expected to increase the salinity level. This will increase the pressure on the available freshwater resources by increasing severe water stress area from 36.4 to 38.6 million km^2. The increase will be significant in Southern Africa, Western Africa and South Asia which will be a limiting factor in the future for industrial and agricultural growth due to competition for water [15]. On an average globally 40% water deficit will be experienced by 2030 under a business-as-usual scenario.

8. Conclusions

In spite of efforts by various stakeholders and global level organizations, lots of gap still persists in our understanding of the global water resources and the emerging salinity problems. Meanwhile there are several disputes in utilizing and sharing this precious resource. Human activities have rendered water unusable at several places due to pollution, salinity and over exploitation. There should be proper regulations and monitoring which involve measures like precaution/prevention; control/restriction and remedial/restoration measures. Efforts should be made to study sea-level rise and sea water intrusion. In summary, the available information suggests that water security and the salinity will remain a challenge for many tropical countries today and in the future until suitable remedial measures are implemented and relevant technologies are developed.

Author details

Ayyam Velmurugan[1*], Palanivel Swarnam[1], Thangavel Subramani[1], Babulal Meena[2] and M.J. Kaledhonkar[2]

1 ICAR-Central Island Agricultural Research Institute, Port Blair, India

2 ICAR-Central Soil Salinity Research Institute, Karnal, India

*Address all correspondence to: vels_21@yahoo.com

References

[1] World Resources Institute. World Resources 2000-2001. Washington, DC: World Resources Institute; 2000. 389 pp

[2] FAO. The use of saline waters for crop production. FAO Irrigation and Drainage Paper 48, Rep. Rome, Italy: Food and Agriculture Organization; 1992. Available from: www.fao.org/3/a-t0667e.pdf [Accessed: 20 September 2018]

[3] Yamaguchi T, Blumwald E. Developing salt-tolerant crop plants: Challenges and opportunities. Trends in Plant Science. 2005;**10**(12):615-620

[4] Mancosu N, Snyder RL, Kyriakakis G, Spano D. Water scarcity and future challenges for food production. Water. 2015;7:975-992. DOI: 10.3390/w7030975

[5] Shiklomanov IA. World water resources and water use: Present assessment and outlook for 2005. In: Rijberman F, editor; World water scenarios analysis; World Water Vision; 2000

[6] FAO. Review of world water resources by country. Water Reports 23. Rome, Italy: Food and Agriculture Organization of the United Nations; 2003. Available from: http://www.fao.org/docrep/005/y4473e/y4473e00.htm

[7] FAO. Climate change, water and food security. FAO Water Reports No. 36. Rome, Italy: FAO; 2011

[8] Siebert J, Burke JM, Faures K, Frenken J, Hoogeveen P, Doll HP, et al. Groundwater use for irrigation—A global inventory. Hydrology and Earth System Sciences. 2010;**14**:1863-1880. DOI: 10.5194/hess-14-1863-2010

[9] Szabolcs I. Global overview of sustainable management of salt-affected soils. In: Proceedings of the International Workshop on Integrated Soil Management for Sustainable use of Salt-Affected Soils; Bureau of Soils and Water Management, Diliman, Quezon City, Manila; 1995. pp. 19-38

[10] Szabolcs I. Salt-Affected Soils. Boca Raton: CRC Press; 1989. p. 274

[11] Shahid SA, Rahman K. Soil salinity development, classification, assessment and management in irrigated agriculture. In: Pessarakli M, editor. Handbook of Plant and Crop Stress. Boca Raton/London/New York: CRC Press/Taylor and Francis Group; 2011. pp. 23-39

[12] Massoud FI. Basic principles for prognosis and monitoring of salinity and sodicity. In: Proceedings of the International Conference on Managing Saline Water for Irrigation; Lubbock, TX: Texas Tech University; 16-20 August 1976; 1977. pp. 432-454

[13] Balba AM. Minimum management programme to combat world desertification. UNDP Consultancy Report on Advances in Soil Water Research; Alexandria, Egypt; 1980

[14] Pessarakli M, Szabolcs I. Soil salinity and sodicity as particular plant/crop stress factors. In: Pessarakli M, editor. Handbook of Plant and Crop Stress. Boca Raton/London/New York: CRC Press/Taylor and Francis Group; 2011. pp. 3-21, 496

[15] Alcamo J, Henrichs T, Thomas R. World water in 2025. Global modeling and scenario analysis for the world commission on water for the 21st century. Centre for Environmental Systems Research, University of Kassel; 2000

[16] IWMI. Water for Food, Water for Life: A Comprehensive Assessment of Water Management in Agriculture. London/Colombo: Earthscan/International Water

Management Institute; 2007. Available from: www.earthscan.co.uk

[17] Lakhdar A, Rabhi M, Ghnaya T, Montemurro F, Jedidi N, Abdelly C. Effectiveness of compost use in salt-affected soil. Journal of Hazardous Materials. 2009;**171**(1-3):29-37

[18] IPCC. Climate change: Synthesis report. Contribution of Working Groups I, II and III to the Fourth Assessment Report of the Intergovernmental Panel on Climate Change. Geneva, Switzerland; 2007. p. 104

[19] FAO. AQUASTAT Database. Food and Agriculture Organization of the United Nations. 2012. Available from: http://www.fao.org/nr/aquastat [Accessed: 20 June 2017]

[20] Alexandratos N, Bruinsma J. World agriculture towards 2030/2050: The 2012 revision, ESA Working Paper No. 12-03; Rome, Italy: FAO; 2012

[21] Hornbogen M, Schultz GA. Water: A Looming Crisis? Paris: UNESCO; 1998. pp. 357-362

Chapter 4

World's Demand for Food and Water: The Consequences of Climate Change

Sheikh Mohammad Fakhrul Islam and Zahurul Karim

Abstract

This study focused on analysis of global food demand and supply situation by 2030 and 2050, water demand-availability, impact of climate change on world water resource, food security and desalination challenges and development opportunities. The population of the world will be 8.6 billion in 2030 and 9.8 billion in 2050; Africa will be the major contributor. World cereal equivalent (CE) food demand is projected to be around 10,094 million tons in 2030 and 14,886 million tons in 2050, while its production is projected to be 10,120 million tons in 2030 and 15,970 million tons in 2050 having a marginal surplus. India and China are capturing large share of global food demand. The developing country will demand more animal origin foods due to income growth in the future. The growth rate of world demand for cereals will decline till 2050. Global water demand is projected to increase by 55% between 2000 and 2050 from 3500 to 5425 km^3. Evidence showed that climate change will have adverse impact on world water resources and food production with high degree of regional variability and scarcity. A number of options are suggested for development of global water resource and food production.

Keywords: demand for food and water, food security, climate change, global, water resources, challenges and opportunities

1. Introduction

Food and water are important for life. Global population increased many folds in the last century and will further boost by 2030 and 2050 [1]. Such large world population will be demanding for more food and water in the future. Despite the fact that agricultural growth has been higher than the rate of population growth concerns has been raised whether the land mass of world is actually capable of supporting its expanding population by 2030 and 2050. Food security remains a relevant and priority of many nations with special emphasis on developing countries. There is growing concern on the future demand for and supply of food in the world. The global food system is experiencing an unprecedented confluence of pressures that may increase over the years 2050 [1]. Increased food production will require greater inputs-land, water or energy, or a combination of these inputs. Thus, required increase in food production will intensify competition for land, water and energy [2, 3]. The global agriculture is evolving with food demand of people, availability of technology and climate change. Could the future growth of

IntechOpen

supply of food of a country match with its increased demand for food as a result of population pressure and rising income? A number of studies attempted to answer this and projected demand for and supply of key food items in various countries and assessed gap [4–6].

There are growing concerns on the impact of climate change on the water resources. A number of studies assessed such impact at various country levels and food security challenges [7]. An ever increasing amount of evidence suggests that the continual increase in greenhouse gas emissions is affecting the global climate and altering the local precipitation and temperatures [8, 9]. Climate change is expected to produce significant effects on global water resources and freshwater ecosystems [10, 11]. The effects and intensity of climate change will vary from region to region [12]. Impact of climate on global water storage capabilities and hydrologic functions will have significant implications for water management and planning as variability in natural processes increases.

This study was carried out with the objectives to examine and assess global food demand and supply situation by 2030 and 2050, world water demand- availability scenario, impact of climate change on global water resource, food security challenges of the globe, identify challenges and development opportunities. The study is completed based on extensive review and analysis of relevant information and literature available across various regions of the globe.

2. Outlook of world population

Population of the world reached to 7.3 billion by mid-2015 and the extent of increase was approximately 1 billion people during the period of last 12 years. The vast majority of the global population (60%) lives in Asia (4.4 billion), the second highest (16%) in Africa (1.2 billion), third portion (10%) in Europe (738 million), the 4th one (9%) in Latin America and the Caribbean (634 million), and the remaining 5% in rest of the world (**Table 1**). China (1.4 billion) and India (1.3 billion) that belong to Asia are the two largest countries of the world, covering 19% and 18 per cent of the world's population, respectively [8].

2.1 Projected growth of population

The growth rate of global population increased slowly during 1700–1950 and then accelerated rapidly until the mid-1960s, peaking at just over 2% per year before descending to 1.1% per year in 2017. World population size increased seven fold during the period 1800–2011.

Currently, the world population is growing approximately by 83 million people annually. Growth rates are slowing to various extents within different regions with

Major region	***Population (millions)***			
	2015	2030	2050	*2100*
World	7 349	8 501	9 725	11 213
Africa	1 186	1 679	2 478	4 387
Asia	4 393	4 923	5 267	4 889
Europe	738	734	707	646
Latin America and the Caribbean	634	721	784	721
Northern America	358	396	433	500
Oceania	39	47	57	71

Table 1.
Population of the world by region.

result of the overall population growth rate decreasing from 1.55% per year in 1995 to 1.10% in 2017. The median estimate for future growth shows that the world population is projected to increase by more than 1 billion people within the next 15 years, reaching 8.6 billion in 2030, further increase to 9.8 billion in 2050 and 11.2 billion by 2100 assuming a continuing decrease in average fertility rate from 2.5 births per woman in 2010–2015 to 2.2 in 2045–2050 and to 2.0 in 2095–2100 (**Figure 1**). With the main driver of future population growth is the evolution of the fertility rate [9].

More than half of global population growth between now and 2050 will occur in Africa. Africa has the highest rate of population growth among major regions, growing at a pace of 2% annually in 2010–2015 (**Figure 2**). An additional 2.4 billion people projected to be added to the global population between 2015 and 2050 of which 1.3 billion will be added from Africa and 0.9 billion people from Asia. Asia is the second largest contributor to future global population growth followed by Northern America, Latin America and the Caribbean and Oceania, which are projected to have much smaller increments. In the medium variant, Europe is projected to have a smaller population in 2050 than in 2015.

There is link of population growth with economic growth and food demand. According to Malthus, population growth responds to a wage or income signal that

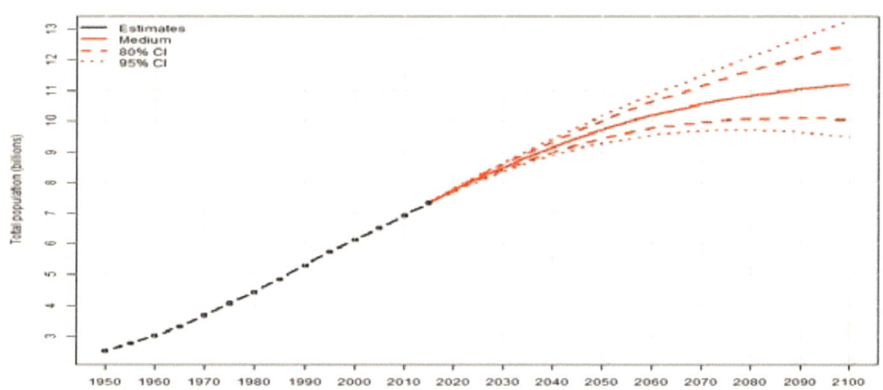

Figure 1.
Median variant projections of world population 2015–2100. Source: Ref. [8].

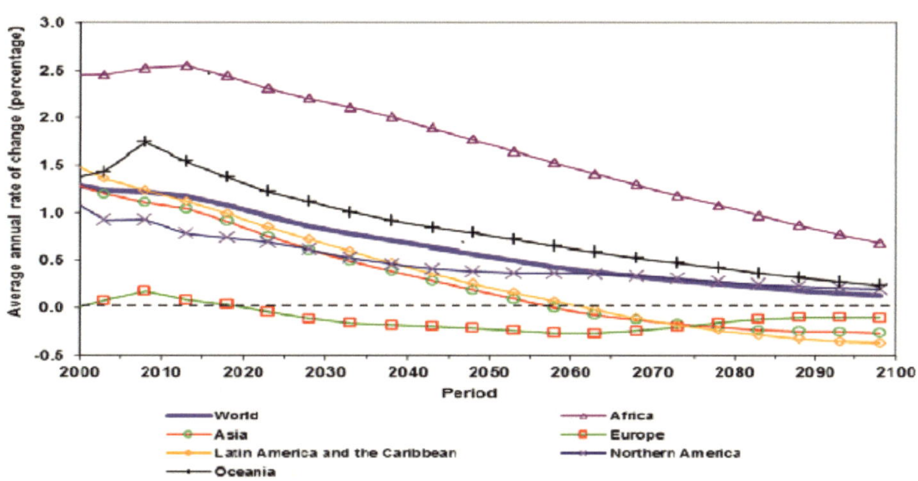

Figure 2.
Medium-variant projection of population growth by major region, 2015–2100. Source: Ref [8].

depends negatively on the size of the population in relation to the economy and its resource base [8]. Population growth is positively related with incidence of poverty. With economic growth incidence of poverty is reduced and population growth declines as result of declining fertility rates. For example, incidence of poverty is high in Africa and growth rate of population is also high.

3. Projections of global food demand and supply

The projected large world population in 2030 and 2050 discussed above has received a great deal of attention as an influence on world food demand [8]. Besides population growth, income growth also becomes an important driver of food demand. According to Bennett's law the proportion of the food budget spent on starchy-staple foods declines while spending on animal-based products increases as incomes grow in developing countries [10]. This dietary change puts pressure on agricultural resources since animal-based food requires disproportionately more agricultural resources including water in production [11]. This relationship between food demand and income, established by Engel's and Bennett's laws, implies that income distribution matters for aggregate food demand.

Substantial efforts have been made in modeling to forecast the global supply and demand for food to the middle of the century, typically using large global agricultural models [12–14]. However, the projections for food output and prices vary widely across the models, depending on their underlying supply and demand specifications, choices of key parameters such as price and income elasticities and their treatments of technical change. For instance, reviewing modeling approaches from 12 global agricultural economic models, It is reported that modelers' projections for increases in global crop output between 2005 and 2050 range from 52 to 116%, while estimated changes in crop prices vary from a decline of 16% to a rise of 46% [15]. Another study projected an increase of 95% in consumption of animal-based food, as against an 18% increase in demand for starchy staples, with the latter being largely driven by population growth toward 2050 [16].

For simplicity of estimation of projected world food demand all food items were converted to cereal equivalent food (CE) [17]. The drivers of cereal equivalent (CE) food demand are growth rate in per capita CE food consumption and population growth rate. **Table 2** shows that world CE food demand increased from 2999 million tons in 1980 to 6360 million tons in 2009. Decade wise analysis of growth rate shows that annual growth rate of CE food demand declined from 2.3% in 1980s to 1.87% during 2001–2009 while per capita food demand increased from 0.55 to

	Evolution of CE food demand 1980-2009					
	CE Food demand			Annual growth rate in CE food demand		
	Change (mil. Tons)	Initial year (mil. Tons)	Last year (mil. Tons)	Total (%)	Per capita (%)	Population (%)
1980-1991	864	2999	3863	2.3	0.55	1.75
1992-2000	817	4590	5407	2.05	0.69	1.36
2001-2009	905	5455	6360	1.87	0.72	1.15
	Projected change in CE food demand 2009-2050					
	CE Food demand			Annual growth rate in CE food demand		
	Change (mil. Tons)	Initial year (mil. Tons)	Final year (mil. Tons)	Total (%)	Per capita (%)	Population (%)
2009-2050	7049	6360	14886	1.72	1.03	0.68

Table 2.
Evolution of world food demand during 1980–2009 and Projections in 2050.

0.72% and population growth rate declined from 1.75 to 1.15% (**Table 2**). World CE food demand is projected under strong convergence scenario to be around 10,094 million tons in 2030 and 14,886 million tons in 2050 [17]. On the supply side, CE food production is projected to be 10,120 million tons in 2030 and 15,970 million tons in 2050. The world CE food demands would change by 134% from the base year of 2009 while CE food production would change by 151% and thus food production would grow slightly faster than demand yielding a positive gap of 7%.

The regional decomposition shows that developing countries as a group dominate the increase in food demand and that their income convergence does matter. It was that convergence by middle-income countries, especially such populous countries as India, China, Indonesia and Nigeria, is particularly important for global food demand. This is partly due to the inverted-U shaped pattern of income elasticities for aggregate food demand, with middle-income countries experiencing the largest income elasticities due to their dietary upgrading toward more resource demanding products [10]. **Table 3** shows top 20 countries contributing around 77.5% of total world CE food demand.

India has the largest share of world food demand (24.3%) followed by China (16.7%). Although Bangladesh is small country but densely populated and stood 8th with a share of 2% of world food demand (**Table 3**). **Table 4** shows projected food demand and supply of Bangladesh by 2030 and 2050. Bangladesh is self-sufficient in rice now. Rice production was 5% less than the demand in 2005 and 2000, but there was a marginal surplus of 5% in 2010.The projections show that Bangladesh will have a surplus rice production of 1.2 million tons (MT) and maize production of 1.8 (MT) by 2030. On the other hand, the country will have deficits productions of wheat, potato, pulses, vegetables, meat, egg and fresh water fish amounting 0.5, 0.8, 0.7, 1.0, 0.1 and 0.7 MT. The country is also highly deficit in well seed production. It is projected that Bangladesh will have a surplus production of rice, maize, potato, vegetable and milk by 2050 and will have deficit production of wheat, pulses, fruits, meat and fresh water fish in 2050.

	Change (million ton)	Share (%)	Cumulative (%)
1. India	1715.9	24.3	24.3
2. China	1176.6	16.7	41.0
3. Nigeria	357.6	5.1	46.1
4. Indonesia	301.0	4.3	50.4
5.United States	241.0	3.4	53.8
6. Pakistan	217.9	3.1	56.9
7. Brazil	178.5	2.5	59.4
8. Bangladesh	140.1	2.0.	61.4
9. Egypt	139.8	2.0	63.4
10. Mexico	133.8	1.9	65.3
11. Philippines	128.8	1.8	67.1
12. Ethiopia	105.1	1.5	68.6
13. Turkey	91.8	1.3	69.9
14. Vietnam	91.2	1.3	71.2
15. Tanzania	81.1	1.2	72.4
16.Russian	76.0	1.1	73.4
17. Sudan	74.5	1.1	74.5
18. Iran	73.5	1.0	75.5
19. Uganda	73.4	1.0	76.6
20. Thailand	67.7	1.0	77.5

Table 3.
Top twenty countries contributing to world CE food demand changes.

Year	Rice	Wheat	Maize	Potato	Pulses	Vegetable	Fruits	Meat	Egg	Milk	Fish
Food supply											
2030	38.7	1.6	2.9	11.8	0.4	6.4	3.6	0.9	0.4	4.3	3.2
2050	46.5	2.1	2.9	18.7	0.4	8.3	3.3	1.2	0.5	5.2	3.9
Food demand											
2030	37.5	4.2	1.1	12.3	1.2	7.0	3.2	1.9	0.5	4.2	3.9
2050	41.3	4.5	1.2	13.3	1.3	7.6	3.5	2.1	0.5	4.6	4.2
Surplus(+)/deficit(-											
2030	1.2	-2.6	1.8	-0.5	-0.8	-0.7	0.4	-1.0	-0.1	0.1	-0.7
2050	5.2	-2.4	1.7	5.4	-0.9	0.7	-0.2	-0.9	0.0	0.7	-0.3

Table 4.
Projections of food supply and demand in Bangladesh by 2030 and 2050 (Based on estimates of ARIMA model, quantity in million tons).

Still cereals constitute major portion of world food demand (49%) and will remain so till 2050. The growth rate of global demand for cereals declined continuously during 1969–2007 from 2% per annum to 1.3% and projected to fall further to 1.2% in 2030 and to 0.9 in 2050 while world cereal demand would have a significant rise from 940 million tons from the base year 2005/2007 to 3 billion metric tons by 2050. Almost all the increases in the consumption of cereals will come from the developing countries. The developing countries surpassed developed ones in total cereals consumption in the early 1980s and account now for 61% of world consumption, a share that will increase to 67% by 2050. They also surpassed them in total production in the early 1990s: they now account for 56% of world production and the share will increase to 60% in 2050 [18].

Like other developing countries with income growth food consumption in Bangladesh is slowly diversifying. Cereals still provide a major part of the calorie intake, but their share in total calorie supply has decreased from 92% in 1990 to 89% by 2010. Auto Regressive Integrated Moving Average (ARIMA) projections show that it will further decrease to 86.6% by 2030 and 85.8 by 2050 (**Figure 2**). The contribution to calorie intake from potato, vegetables, and animal products gradually increased between 1990 and 2010 and will continue to increase between 2030 and 2050 (**Figure 3**). The share of rice will decrease from 82% in 2010 to 79% in 2030 and to 78.6% in 2050 and absolute consumption decrease by 24.5 kcal/person/day from 2010 level (**Figure 3**). The share of wheat will slightly decrease from 7% in 2010 to 6.8% in 2030 and 6.7% in 2050 and absolute consumption decrease by15.1 kcal/person/day (**Figure 3**). The share of calorie intake from cereals seems to be reaching a level of saturation. However, as far as rice consumption is concerned, there is no room for significant increases in average consumption even with income

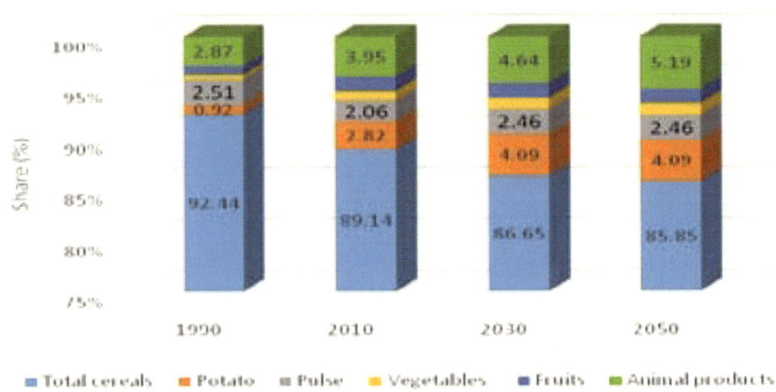

Figure 3.
Share of major food items in total calorie intake per capita. Source: Author's estimation.

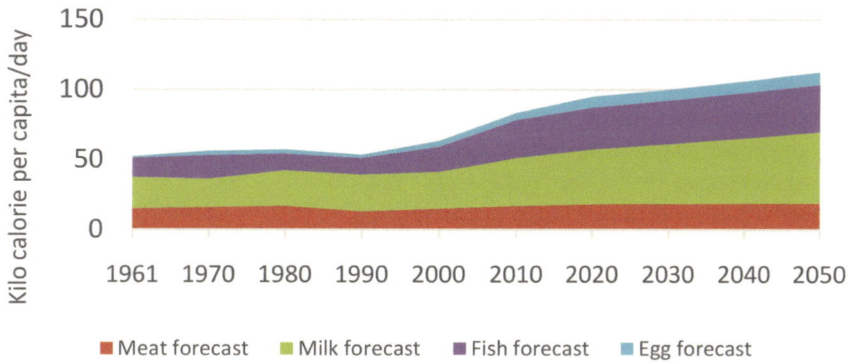

Figure 4.
Projection of per capita calorie intake from animal products. Source: Author's estimation, Per capita calorie intake up to 2013: FAOSTAT and forecasts up to 2050 are authors' estimates.

growth; in fact, it even started decreasing as in countries with similar consumption and economic growth patterns in Asia [19].

World average per capita rice consumption has declined after late 1980s, following mild declines in several countries of East and South Asia and small increases in other regions. These trends are projected to continue and the average of the developing countries may fall from the present 64 to 57 kg in 2050 [18]. It is striking to note that the per capita wheat consumption has also declined in both the developing and the developed countries. Food consumption demand of coarse grains as staple food in several countries in sub-Saharan Africa will increase in the next decades.

With the growth in income in developing countries demand for food from livestock origin increased in the past two decades and projected to grow further by 2030 and 2050. The ARIMA forecasts show that the consumption of animal origin food (meat, milk, egg and fish) and non-cereal food (potato, vegetables and fruits) in Bangladesh will have increasing trend during 1990–2030 (**Figure 3**). Beyond 2030 the consumption of animal products will further increase (**Figure 4**).

4. Global water demand-availability analysis for 2030 and 2050

The demand for water originates from four main uses- agriculture, energy production, industrial uses and human consumption. Production of crops and livestock is water-intensive as a result agriculture is the largest water user accounting 70% of global water withdrawal and rest 30% is used by municipal, energy and industrial sectors. The global booming demand for livestock products is increasing the demand for water as well. The global demand for food is expected to increase by 70% by 2050 [20]. Over the past half century, the area equipped for irrigation has more than doubled, total livestock production has more than tripled and inland aquaculture has grown more than 20-fold.

Global water demand is projected increase by 55% between 2000 and 2050 from 3500 to 5425 km^3 (**Figure 5**). In addition to demand from the agricultural sector, large increases are predicted for industry (400%), energy production (140%) and domestic use (130%) [21, 22]. Accelerated urbanization and the expansion of municipal water supply and sanitation systems would also contribute to the rising demand. Changing consumption patterns, including shifting diets toward highly water-intensive foods such as meat (i.e. 15,000 liters of water are needed for 1 kg of beef) will worsen the situation. While a person may drink 2–4 liters of water a day, it takes 2000–5000 liters of water to produce a person's daily food. Water is important for food security, crop growth, livestock, and food markets. Lack of water can be a major cause of famine and undernourishment, especially in areas where people depend on local agriculture for food and livelihoods. OECD projected that 3.9 billion people - in total over 40% of the world's population - are likely to be living in river basins under severe water stress by 2050 (**Figure 6**). Near East/North Africa and Northern China are water scarce regions.

The world net-land under crops predicted to increase by some 70 million ha by 2050. The area harvested may increase by almost twice that amount as a result of increased multiple cropping and reduced fallow lands. The projected 70 million ha increase is the result of an expansion in the countries of sub Saharan Africa and Latin America [18]. Irrigation has been an important contributor to cereal yield growth over the past decades. World irrigated areas are currently 300 million ha, more than twice the level of the early 1960s. World total irrigated area is projected to expand to 322 million ha in 2050. This expansion of around 22 million ha will be mainly in developing countries. The potential for further expansion of irrigation is limited.

Many water sources of the world are degrading and creating water scarcities. Most of the world irrigated agriculture is today in developing countries, accounting 60% of their cereals production. Nearly one half of the irrigated area of the developing countries is in India and China. One third of the projected increase will likely be in these two countries. The renewable water resources that would underpin the expansion of irrigation are extremely scarce in several countries. Irrigation water withdrawals from such resources are only 6.6% globally and even less in some regions. However, in the Near East/North Africa and in South Asia they already account for 52 and 40%, respectively, in 2005/2007. For some countries of Central America and the Caribbean these percentages are higher. Any country using more

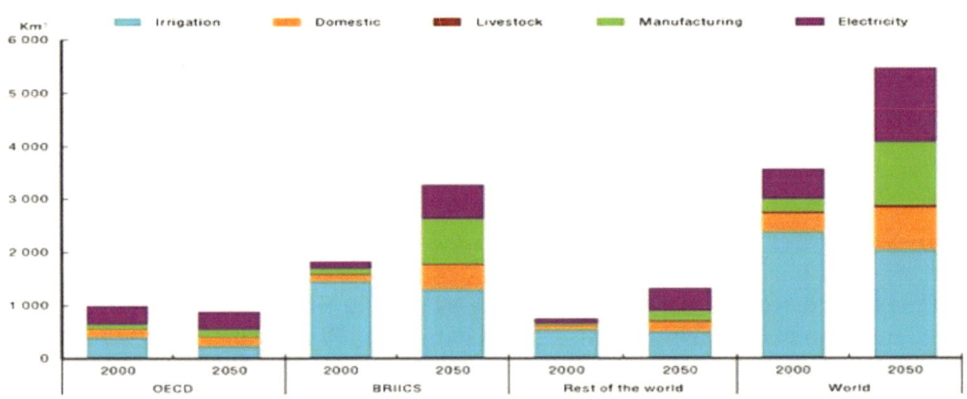

Figure 5.
Projection world water demand in 2050. Source: Ref. [22].

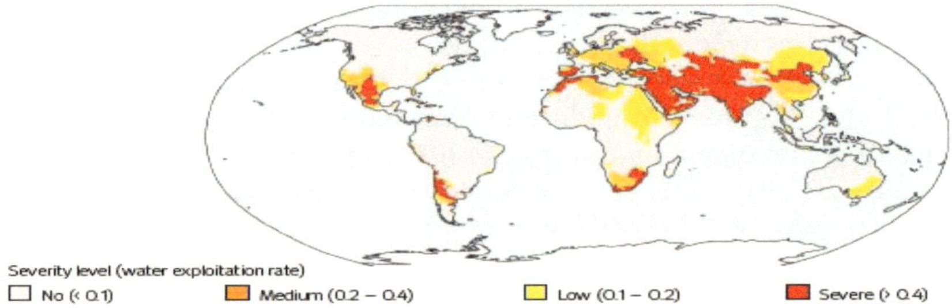

Figure 6.
Population projected to living in river basin under severe water stress. Source: Ref. [22].

than 20% of its renewable resources for irrigation is considered as crossing the threshold of impending water scarcity. There are already 22 countries (developing but including some in the Central Asia region) that have crossed this threshold, 13 of them in the critical over 40% class. Libya, Saudi Arabia, Yemen and Egypt use volumes of water for irrigation larger than their annual renewable resources [18].

Rice production accounted for 93% of the total consumptive water use (CWU) and 90% of the total irrigation CWU in Bangladesh in 2015. Boro rice alone occupies the largest share of irrigation water. We have projected that water demand for Boro rice would increase from 16.5 Billion meter^3 (BM^3) in 2010 to 17.23 BM^3 in 2030 after that it will stabilize and would remain at 17.23 BM^3 in 2050 (**Table 5**). Surface water availability in Bangladesh is alarmingly declining due to more water withdrawal in the upper riparian countries, silting up of major rivers and adverse impact of climate change. Salinity front in the south is also penetrating more inlands of the country due to shrinking of surface water. Thus the groundwater is the major source for more than 75% of the irrigated area in Bangladesh, it amounts 13 BM^{3}of irrigation water in 2010. The projected 17.23 BM^3 water for Boro rice production in 2030 mostly would come from groundwater.

The sources of water in Bangladesh can be classified as surface water, rainfall and ground water. Bangladesh, being the lower most riparian country in the Ganges-the Brahmaputra-the Meghna basins and crisscrossed by around 700 rivers including 57 transboundary rivers, shares its trans-boundary water resources with the upper riparian countries like Bhutan, China, India and Nepal. In the past few decades reduction of dry season flows in Bangladesh due to increasing upstream withdrawal is causing severe water shortage across the country [23]. For instance, due to withdrawal of water from the transboundary Teesta River through construction of multi-purpose barrage and dams by the upper riparian countries, water availability in Bangladesh portion of the river gradually reduced to 6500 cusec in 1997 and it drastically reduced over the years to 250 cusec in 2015 against Bangladesh's requirements of 8000 cusec. Teesta already silted much due to low flow of the river and its branches of many small rivers have dried up. Moreover, the reduced stream flow is also accelerating salinity intrusion and environmental degradation, particularly in the South West region. Again, excessive release of water from upper catchment during monsoon season causes flooding and river bank erosion in Bangladesh [24].

Already there has been much stress on ground water level of Bangladesh due to excessive withdrawal. Given the falling groundwater tables and water quality issues in Bangladesh, it will be extremely difficult to exploit groundwater resources sustainably to meet projected demand. Evidence showed that some districts of North, South and Central regions of Bangladesh already crossed the sustainable thresholds of groundwater use. Groundwater withdrawals for irrigation in these regions may exceed the usable recharge limit [4].

Global water scarcity is growing severe recent years. Recent research has demonstrated that two-thirds of the world's populations currently live in areas that experience water scarcity for at least 1 month a year. About 50% of the people facing this level of water scarcity live in China and India. About 500 million people live

Year	Boro irrigated area (million ha)	Water demand (CWU in billion meter^3)
2010	4.5	16.5
2030	4.7	17.23
2050	4.7	17.23

Table 5.
Irrigation water demand for rice production in Bangladesh in 2030 and 2050.

in areas where water consumption exceeds the locally renewable water resources This includes parts of India, China, the Mediterranean region and the Middle East, Central Asia, arid parts of Sub-Saharan Africa, Australia, Central and Western South America, and Central and Western North America. In these regions groundwater continue to decrease and become highly vulnerable [25].

The availability of water resources is inherently linked to water quality. The pollution of surface water and groundwater may prohibit its different uses due to absence of pre-treatment. The deterioration of water quality is expected to increase further in the coming decades which will further endanger human health and the environment as well as constraining sustainable economic development. The release of untreated wastewater from expanding human settlements and increasing industrial production generates physical, chemical and biological pollution that negatively impact human health and ecosystem. Findings from the global water quality monitoring program showed that severe pathogen pollution affects around one third of all river surface waters in Africa, Asia and Latin America, putting the health of millions of people at risk [26].

Intensive use of fertilizers, agrochemicals and animal waste can accelerate the eutrophication of freshwater and coastal marine ecosystems and increase groundwater pollution. Most of the largest lakes in Latin America and Africa have seen increasing anthropogenic loads of phosphorus, which can accelerate eutrophication processes. Increased discharges of inadequately treated wastewater, resulting from economic and industrial development, intensification and expansion of agriculture, and growing volumes of sewage from rapidly urbanizing areas are contributing to the further degradation of water quality in surface and groundwater around the world. As water pollution critically affects water availability, it needs to be properly managed in order to mitigate the impacts of increasing water scarcity [26].

The municipal and industrial wastewater treatment in high-income, upper middle-income, lower middle-income and low-income countries are about 70, 38, 28 and 8%, respectively. Globally over 80% of all wastewater is discharged without treatment. In high-income countries, the motivation for advanced wastewater treatment is either to maintain environmental quality, or to provide an alternative water source when coping with water scarcity. Recently, the situation of water security in the most populous and rapidly developing mega cities of Asia is worsening because of major challenges resulting from overexploitation of groundwater, skewed water supply and demand due to population explosion and negative impacts of climate change [27].

In addition to hydrologic and climatic impacts, the non-climatic drivers of freshwater systems are changes in population, food demand, economic growth, technology, living standard and societal values of freshwater ecosystems. Land use change, construction and management of reservoirs, pollutant emissions, water treatment and water management influence availability and quality of freshwater at the national and international level.

5. Impact of global climate change on water resources

A large volume of data base is now available on impact of climate change of global water resources. According to IPCC World temperature, humidity and precipitation will change significantly by 2030 and 3050 due to climate change [36]. The variations in the changes in precipitation in a warming is complex across the regions due to climate change will not be uniform (**Figure** 7). The high latitudes and the equatorial Pacific are likely to experience an increase in annual mean precipitation by the end of this century. In many mid-latitude and subtropical dry regions, mean precipitation will likely decrease, while in many mid-latitude wet regions,

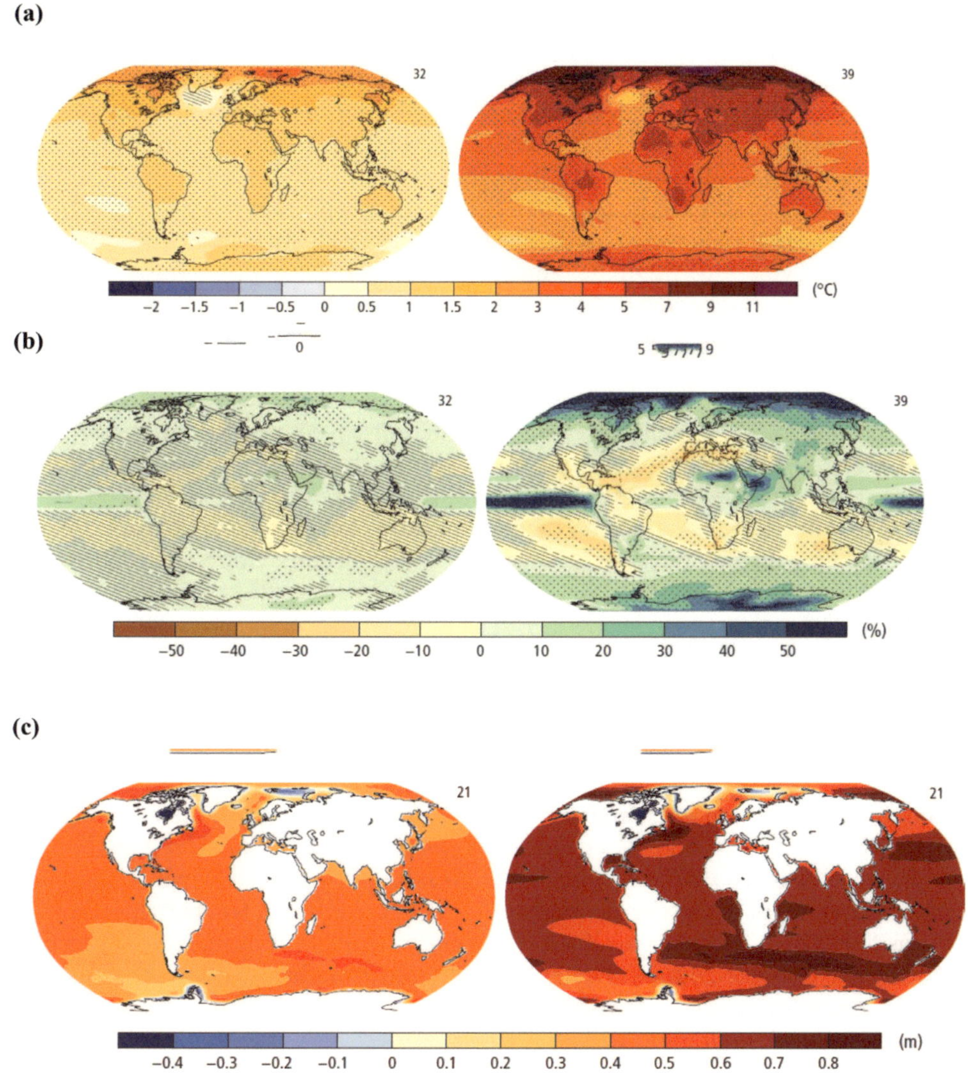

Figure 7.
Projections for the 2081–2100 period under the scenarios for figure (a) change in annual mean surface temperature and figure (b) change in annual mean precipitation, in percentages, and figure (c) change in average sea level. Changes are shown relative to the 1986–2005 period. Source [36].

mean precipitation will likely increase. Extreme precipitation events over most mid-latitude land masses and over wet tropical regions will very likely become more intense and more frequent as global mean surface temperature increases. Globally, it is likely that the area encompassed by monsoon systems will increase and monsoon precipitation is likely to intensify and regional variability [28]. Analysis of historical time series on the occurrence of hundred-year floods in large-scale river basins around the world with the relationship of climate change showed that intensity of flood due to climate change will continue to grow in the future [29].

Continuous increase in greenhouse gas emissions is affecting the global climate that altering the local precipitation, temperatures and atmospheric composition [29, 30]. The global temperature increased by 0.85°C during 1880–2012, and will further increase by 0.3–4.8°C until 2100 [31]. Such global warming will produce significant effects on water resources and freshwater ecosystems [31, 32]. The effects and intensity of climate change will vary from region to region [33].

5.1 Regional variability

Asia: Arid and semi-arid regions in northwestern China are very vulnerable to the effects of climate change due to their fragile ecosystems and scare water resources [34–36]. This region characteristically experiences an extremely dry desert climate with low amounts of infrequent rainfall and strong potential evaporation [34]. Many studies indicated that this region is especially sensitive to climate change because the main water sources in this region come from high elevation glaciers and snowmelt through the largest inland Tarim River which flows through the arid and semi-arid region of northwestern China [33].

The effects of climate change on water resources of the Hotan River Basin in Xinjiang of China was assessed using hydrological models to evaluate responses of discharge, extreme events, evapotranspiration and snowmelt accumulation with the effects of changing climate [37, 38]. The precipitation is projected to experience an overall increase with rates ranging −1.2 to 32.7%. The dry season is predicted to have relatively higher increases than the wet season while a slightly decreasing trend was predicted for July (August and September). The projected average temperature was expected to increase by 1.60–2.61°C. The projected maximum temperature increased slightly during summer and autumn, which represents the predicted warmer daytime temperatures. Discharge will increase with an increase of precipitation. With an increase in temperature, the discharge significantly decreased. The evapotranspiration rate will increase significantly by 7.4–31.3%. Climate change is predicted to lead to stronger changes in peak flow. Stream flow is generally predicted to increase, while the shrinking of snow storage and a reduction in the snowpack will sharply reduce the solid water storage capacity of the landscape. The increasing frequency of extreme events and a spatiotemporal redistribution of water resources will produce great challenges related to agricultural water allocation and management in this region.

Climate change could have a significant impact on drought in North Korea. Drought characteristics in the Hwanghae Plain of North Korea were analyzed from 1981 to 2100 [28]. The results indicated that severe drought is more likely to occur in future as a result of climate change. The seasonal drought conditions were also significantly influenced by climate change.

In a high density populous country like Bangladesh, the effects of climate change on the surface and ground water resources is severe. Changes to water resources and hydrology could lead to adverse impact on the country's economy, where the population is dependent on the surface water for irrigation, industrial production, navigation and various other activities.

Water resources of Bangladesh would be severely affected due to adverse impact of climate change will the most critical for Bangladesh – largely related to coastal and riverine flooding and also enhanced possibility of winter (dry season) drought in northern areas. Both coastal flooding (from sea and river water), and inland flooding (river/rain water) are expected to increase. Flood prone area constitutes about 30% of the land mass and is spread throughout the country. The areas adjacent to major rivers and chars or riverine islands are expected to experience higher intensity flooding. Droughts will be prevalent in the north-west zone of Bangladesh and predicted to reach out into the mid-western region and in the south Cyclones, floods, coastal erosion, and salinity problems may intensify and become more frequent in the 19 districts situated in the coastal zone of Bangladesh. Salinity intrusion from the Bay of Bengal already penetrates 100 kilometers inside the country during the dry season while climate change in its gradual process is likely to further deteriorate the existing scenario. There are 13% areas are with salinity in the southwestern coastal districts of Bangladesh at present, which will increase 16% in 2050 and 18% in 2100 [23].

Europe: The main climate change consequences in Europe related to water resources are increases in temperature, shifts in precipitation patterns and snow cover, and a likely increase in the frequency of flooding and droughts.

Depending on the region, climate change will have widely differing effects on Europe's water. Higher temperatures will generally intensify the global hydrological cycle. Annual precipitation trends in Europe indicate that northern Europe has become 10–40% wetter over the last century, whereas southern Europe has become up to 20% drier. Over the last century annual river discharge increased in some regions, such as Eastern Europe, while it has fallen in others, such as southern Europe.

Climate change may also markedly change the seasonal variation in river-flow. Higher temperatures will push the snow limit upwards in northern Europe and in mountainous regions. This, in conjunction with less precipitation falling as snow, will result in a higher winter run-off in northern European and mountain-fed rivers. Moreover, earlier spring melts will lead to a shift in peak flow levels. As a result of the declining snow reservoir and decreasing glaciers, there will be less water to compensate for the low flow rates in summer.

Climate change tends to increase the frequency and intensity of rainfall; there may be an increase in the occurrence of flooding due to heavy rainfall events. Groundwater recharge may also be affected with a reduction in the availability of groundwater for drinking water in some regions.

In the long-term most climate change scenarios predict that northern and Eastern Europe will see an increase in annual average river flow and water availability. In contrast, average run-off in southern European rivers is projected to decrease. In particular, some river basins in the Mediterranean region, which already face water stress, may see marked decreases of water availability.

The change in temperature is generally more pronounced in higher latitudes, and the air temperature over the European continent has warmed more than the global average, with a 0.8–0.95°C increase since 1900. Important are the regional characteristics of temperature change: the warming has been greatest in Northwest Russia, northern Scandinavia and western Mediterranean. Other parts of Europe, especially central Europe and the eastern Mediterranean coast, show lower increases in temperature or even some decreases (Southeast Germany, Northeast Italy, Macedonia and northern Greece).

The observed higher temperatures stimulate the global hydrological cycle (more evapotranspiration leads to more water vapor in the atmosphere and to more precipitation). Consequently, the average atmospheric water vapor content has increased since at least the 1980s over 10 land and ocean as well as in the upper troposphere [39]. Large areas in the Mediterranean region and in central and eastern Europe experienced a decrease in precipitation over the last century. The observed precipitation trends for the period from 1900 to 2000 show a contrasting picture between increases in northern Europe by 10–40%, and decrease in southern Europe with up to 20% less precipitation, especially in the winter season [40, 41].

River discharge decreased considerably in some southern European river basins and increased in some rivers of Eastern Europe. Extreme floods occurred during the last decade in Germany, Austria, the Czech Republic, Hungary and Poland. A long-term trend toward shorter duration of ice cover has been reported for lakes in Finland and Switzerland. Changes in Europe's water resource will have consequences for several economic sectors. Low water and droughts have severe consequences on most sectors, particularly agriculture, forestry, energy, and drinking water provision. Moreover, wetlands and aquatic ecosystems will be threatened.

Africa*:* The major effects experienced in different African countries are summarized in **Table 6**. Countries in sub-Saharan Africa are likely to suffer the most devastating impacts of climate change. Effects of climate change on water resources in Africa include: flooding, drought, change in the frequency and distribution of rainfall, drying-up of rivers, melting of glaciers, receding of water bodies, landslides, and cyclones among others. Much of Africa is vulnerable to flooding: flood is the most prevalent disaster in North Africa, the second most common in East, South and Central Africa, and the third most common in West Africa [42].

Latin America and Caribbean: Climate change is an important agenda in Central America. This region, together with the Caribbean, is highly vulnerable to the effects of climate change in Latin America. Climate change is manifesting itself through higher average temperatures and more frequent droughts that result in higher water stress, and through the rising frequency of extreme weather events such as tropical storms, hurricanes, floods and landslides, all of which pose significant challenges in the water supply and sanitation sector [43].

Results showed that the regional 16 countries of South America from North West, Central, Northern, Southern and NW Central regions could experience a range of runoff changes depending on whether and how climate change affects precipitation and temperature patterns over the continent [44]. The water availability in the region will be negatively affected by climate change in the next century. Climate change impact assessment indicates that water availability, as reflected by the projected water balance, will likely decrease in most of Nicaragua's basins [45, 46]. The three future scenarios analyzed earlier are in agreement that by 2050 the water

Country	Climate change related effects
Algeria	Mountainous areas subject to severe earthquakes; mudslides and floods in rainy season
Angola	Locally heavy rainfall causes periodic flooding on the plateau
Botswana	Periodic droughts; seasonal August winds blow from the west, carrying sand and dust across the country, which can obscure visibility
Burkina Faso	Recurring droughts
Burundi	Flooding, landslides, drought
Cameroon	Volcanic activity with periodic releases of poisonous gases from Lake Nyos and Lake Monoun volcanoes
Central African Republic	Hot, dry, dusty winds affect northern areas; floods are common
Chad	Hot, dry, dusty winds occur in the north; periodic droughts; locust plagues
Cote d'Ivoire	Coast has heavy surf and no natural harbors; during the rainy season torrential flooding is
Democratic Rep. of the Congo	Periodic droughts in the south; Congo River floods (seasonal); in the east, in the Great Rift Valley, there are active volcanoes
Egypt	Periodic droughts; frequent earthquakes, flash floods, landslides; hot, driving windstorm occurs in spring; dust storms, sandstorms
Ethiopia	Geologically active Great Rift Valley susceptible to earthquakes, volcanic eruptions; frequent droughts
Gambia	Drought (rainfall has dropped by 30% in the last 30 years)
Ghana	Dry, dusty, northeastern winds occur from January to March; droughts
Glorioso Islands	Periodic cyclones
Kenya	Recurring drought; flooding during rainy seasons
Mali	Hot, dust-laden haze is common during dry seasons; recurring droughts; occasional Niger River flooding
Morocco	Northern mountains geologically unstable and subject to earthquakes; periodic droughts
Mozambique	Floods
Namibia	Prolonged periods of drought
Niger	Recurring droughts
Nigeria	Periodic droughts; flooding
Senegal	Lowlands seasonally flooded; periodic droughts
Somalia	Recurring droughts; frequent dust storms over eastern plains in summer; floods during rainy season
South Africa	Prolonged droughts, floods
Sudan	Dust storms and periodic persistent droughts
Swaziland	Drought
Tanzania	Flooding on the central plateau and south-eastern coastal areas during the rainy season;
Zimbabwe	Recurring droughts; floods and severe storms are rare

Table 6.
Climate change-related effects on water resources in Africa.

balance will be reduced in many areas of the country. A reduction in surface water will cause a reduction in groundwater levels and the amount of water available for agriculture, potable water supply and other uses. Current flood-prone areas on Nicaragua's Pacific and Atlantic coasts will likely be exposed to higher runoff than what they are experiencing today.

North America: Climate change is expected to alter hydrologic processes in the Pacific Northwest region of North America, thereby affecting key resources and processes including water supply, infrastructure, aquatic habitat, and access. A warmer climate will affect the amount, timing, and type of precipitation, and the timing and rate of snowmelt which will in turn affect snowpack volume, stream temperature. Altered precipitation patterns would also affect vegetation which would in turn affect water supply [47–50]. There is some indication of increased drought severity and duration in the western and southwestern United States. There is a trend toward reduced mountain snowpack and earlier spring snowmelt runoff peaks across much of the western United States. This trend is very likely attributable at least in part to long-term warming may have substantial impacts on the performance of reservoir systems.

Australia: Plenty of studies have been carried out on the quantitative analysis of the influence of climate change on the hydrological processes. Analysis showed that in 22 basins in Australia will have a change of precipitation and potential evapotranspiration by 1% would cause a change of runoff by 2.1–2.5% and 0.5–1.0%, respectively [51]. Water quality is sensitive to both increased water temperatures and changes in precipitation.

Thus, it appeared from above discussions that climate change will change the world of the present situation of the hydrologic cycle, and cause the redistribution of water resources in time and space.

6. Challenges

6.1 Food security challenges in the globe

The prevalence of undernourishment (POU) in the globe declined considerably in the past decades and reached to 10.9% in 2017. It is projected that POU will have increasing trend beyond 2017 due to persistent conflicts in regions, adverse climate events and economic slowdowns that had affected more peaceful settings and worsened the food security situation. Evidence confirms that lower levels of per capita food consumption in some countries and increased inequality in the ability to access food in the populations of developing countries are contributing to increasing trend in POU [52, 53].

Africa has highest proportion of population (20.4%) having suffering from PoU (more than 256 million people). The prevalence of undernourishment in Africa and Oceania has been increasing for a number of years. This trend is observed in all sub regions of sub-Saharan Africa except for Eastern Africa. A further slight increase is seen in Southern Africa, while a significant uptick is seen in Western Africa, possibly reflecting factors such as droughts, rising foods prices and a slowdown of real per capita Gross Domestic Product (GDP) growth (**Table 7**).

Asia has the highest number of people undernourished (515 million, 11.4% of population). Although In the past decades Asia had decreasing trend in POU until recently it is ended now. Western and South-eastern Asia are among those contributing to this slowdown in the decreasing trend, reflecting the fact that countries in South-eastern Asia have been affected by adverse climate conditions with impacts on food availability and prices, while countries in Western Asia have been affected by prolonged armed conflicts [52].

	Prevalence of undernourishment (%)					
	2005	2010	2012	2014	2016	2017
WORLD	14.5	11.8	11.3	10.7	10.8	10.9
AFRICA	21.2	19.1	18.6	18.3	19.7	20.4
ASIA	17.3	13.6	12.9	12.0	11.5	11.4
LATIN AMERICA AND THE CARIBBEAN	9.1	6.8	6.4	6.2	6.1	6.1
OCEANIA	5.5	5.2	5.4	5.9	6.6	7.0
NORTHERN AMERICA AND EUROPE	< 2.5	< 2.5	< 2.5	< 2.5	< 2.5	< 2.5

Table 7.
Prevalence of undernourishment in the world.

South America has relatively low level of undernourishment, and the situation on POU is deteriorating. POU has increased from 4.7% in 2014 to a projected 5.0% in 2017.

The growth in global food production was higher than the population growth due to adoption of high yielding variety seed, fertilizer and irrigation technology (**Figure 8**). Over the past 50 years, the amount of food available per person has increased by 20%. During the second half of the twentieth century, global food availability and access developed rapidly enough to supersede population growth. As a result, many countries improved their food security and made impressive achievements in reducing hunger and malnutrition by 2015. With the existing technology it will be difficult to boost food production further in the future, specifically during 2030 and 2050. Breakthrough in technology should include new varieties of (rice and wheat) with much high yield ceiling, efficient resource management, faster mechanization and developing high skilled farmers with wider employments of women would be need to transform global agriculture to feed the increased population in the coming decades.

Promoting sustainable agricultural productivity growth is the key to ensuring food availability at affordable prices. While it is likely to become increasingly difficult to push yield frontiers at a constant percentage rate of growth, there is great scope for developing countries to close the yield gap between actual and potential. There is much less scope for increasing cultivated land area of the world.

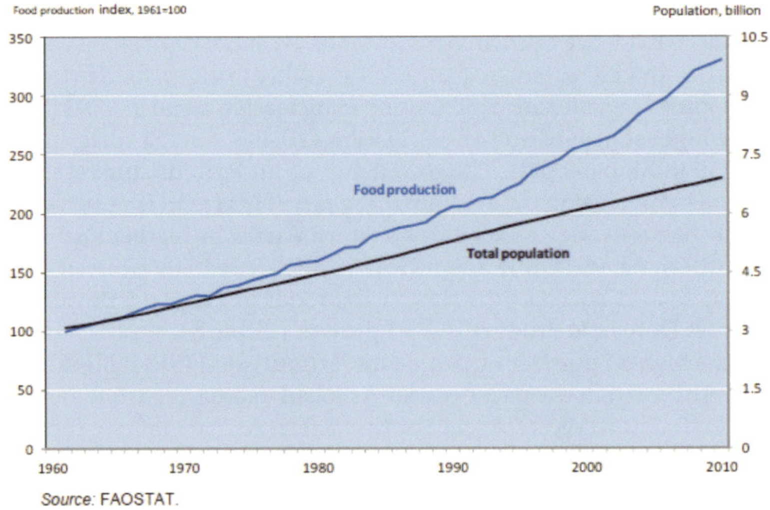

Figure 8.
Global food production and population growth.

So, it is necessary to improve yields of food grains rather than expanding cultivated area. Moreover, a large share of the world's agricultural production is based on the unsustainable exploitation of water resources. There is a need for policies to manage both land and water resources sustainably.

In the coming decades food security threatened due to the fact that climate variability and extremes are negatively affecting agricultural productivity globally. Rising temperature and variability in precipitation would bring changes in global cropping areas, cropping intensity and crop yields. A number of studies shown evidences that both cropping intensity and cropped areas are negatively affected by climate variations and extremes. For example, in the Viet Nam Mekong Delta, variations in the timing and extent of flooding in the wet season and salinity intrusion in the dry season are affecting rice cropping cycles. Recent occurrence of severe floods in Bangladesh in 2018 led to failure of Boro rice crop and threatened its food security. Of course, climate impacts vary between regions, countries, and within a given country due to the diversity and complexity of agricultural systems.

Crop yields in many countries have suffered from changes in temperature and precipitation, which have affected global aggregate rice, wheat and maize yields. A number of studies indicated that heat and water stress resulting in significant global inter annual variability of yields for rice, wheat and maize. Global Synthesis of 144 studies across all regions showed that yield of maize and wheat could be reduced by 20.6 and 39.3% due to drought [54]. Evidences shows that yields of rice, wheat and maize will be declined by a significant amount (roughly one fourth) toward the end of this century due to climate factors (**Box 1**). IPCC Fifth assessment report projected a negative yield impacts for all crops for 3°C of local warming without adaptation across the globe, even with benefits of higher CO_2 and rainfall. South Asia and southern Africa in the absence of adaptation, would suffer the most negative impacts on several important crops [7].

The existence of large numbers of undernourished people is correlated with reduced yields due to increased climate variability and extremes. For instance sub-Saharan Africa has a high level of undernourished people, a region that already has the lowest crop yields globally; increasing temperatures reduced yields for

By the end of this century, the average global temperature is predicted to rise due to the increasing release of greenhouse gases into the atmosphere. Different predictive models inferred that climate change would reduce yields of major cereal crops across different regions of the globe due to rising temperature, resulting in food insecurity challenges.

Rice

Various researchers have shown that global warming can have a negative impact on the yields of paddy produced around the world [55]. It is projected that paddy yield will decrease by 10–15% [56, 57]. In Malaysia that a 2°C temperature increase could reduce paddy yields by 0.36 t ha^{-1} [58]. Temperature increase more than 4°C reduced rice yield in Bangladesh by 36% [59].

Maize

Maize yield in Malawi will decrease 14% by mid-century, and 33% by the century's end because of climate change, in China in maize yield will decline by 35% in 2030 [60], in USA corn yields are projected to decrease further by 20–50% by 2050 [61], in Africa maize yield will decrease by 20% [62], In France, USA, Brazil, and Tanzania, maize yields reduced by about 0.5 t ha^{-1}per 1°C raise in temperature [63].

Wheat

Global climate changes and extreme weather events will have a huge impact on the production of wheat, one of the most widely consumed cereals. In France wheat yield would be reduced by 3.5–12.9% in the medium term from 2037 to 2065, it will further decline by 14.6–17.2% by the end of the century [64]. In China, researchers reported that wheat production rates would be reduced by 3–10% due to a 1°C [65]. In Turkey wheat production would decline by 8–23% by the end of 2100. In Bangladesh 2% increase in winter season temperature wheat yield will reduce by 20 [59] and 4% increase in temperature reduced its yield by 31% [59].

Box 1.
Reduction in cereal yields due to climate change.

maize, sorghum and groundnuts. In semi-arid climate regions such as Central Asia, the Near East, and Northern Africa, cereal production is also highly dependent on climate variability Drought is one of the most important climate events that have been shown to have a negative impact on production. For many countries, there is a high negative correlation between drought indicators and food production. The highest correlations occur in semi-arid countries or drought-prone continental climates (example. Central Asia). In rural India, higher numbers of hot days during the agricultural season are leading to lower crop yields. The impact of drought on decreasing crop yields is widely documented [58].

Apart from production side, post-harvest loss in food is huge and reducing food availability which accounting around one-third of all production in developing countries. Food availability could be enhanced and made sustainable through reducing post-harvest loss with increased investments in market infrastructure, value addition and food processing.

The principal cause of food insecurity is poverty and inadequate incomes. Although globally there is enough food available but many people are too poor to afford it. Tighter world food markets could not quickly respond to supply shocks due to natural calamities causing less food available associated with a price hike reducing affordability of the poor people. For instance incidence of devastating cyclone Sidr in 2007 and Aila in 2009 caused food shortage of Bangladesh and created soaring food prices that hit hard the poor people. Therefore, broad-based income growth is the key to lasting reductions in global hunger. Moreover, promoting international trade could contribute much toward global food availability. Food deficit countries would be able to import food from the surplus countries. Functioning of flexible world food markets would reduce volatility in food prices and consumers will be benefited during food crisis with ample supply at affordable price. Moreover, export promotion would increase income of the small farmers producing exportable fresh commodities. For instance large numbers of small farmers are linked in production system of fresh vegetables, fruits and shrimp in Bangladesh and export chains. The small farmers are getting higher prices with the promotion of export of these produces and their income and purchasing power are improving.

6.2 Challenges of desalination

We have discussed earlier about scarcity of fresh water in the globe. Only 2.5% of the Earth's water is fresh non-salty and major portion of it is ice and glaciers (97.2%) contained within the Earth's Polar Regions. In addition, another 1.8% of that exists below ground in the form of underground rivers and aquifers. This means that the amount of water that exists as groundwater, rivers, lakes, and streams which is immediately accessible for drinking and irrigation is just 0.7%. The remaining 97.5% is salt water available from oceans.

Because of population growth, industrialization and climate change, water scarcity has become one of the most pervasive problems afflicting people throughout the world. Presently, over one-third of the world's population lacks access to safe drinking water and suffers the consequences of unacceptable sanitary conditions [66, 67]. According to the International Desalination Association (IDA), in June 2011, 15,988 desalination plants operated worldwide, producing 66.5 million cubic meters per day, providing water for 300 million people. However, the vast majority of this production took place within countries where access to freshwater is limited and cheaper alternatives (such as drilling for aquifers) are not available.

The researchers have been seeking cost-effective ways of turning sea water into drinking water for decades. Development continued and in the 1970s, commercial membrane processes - such as reverse osmosis (RO) and electro dialysis (ED) - began to be used more extensively. Since 1980 reverse osmosis (RO) desalination technology is commercially used in regions and municipalities all around the world where fresh water supplies were limited. At present, reverse osmosis (RO) accounts for approximately 60% of installed capacity. Desalination is an energy intensive technology, and its future costs will depend much on the price of energy Hence it is necessary to develop more cost-effective processes and use renewable energy for desalination plants.

Environmental considerations are also a going concern for desalination of water plants. There was a rapid growth in the installation of brackish water reverse osmosis (BWRO) desalination facilities in the past decade. Nations, spanning from Australia to Spain, from the United States to China, all have BWRO desalination projects accomplished and construction of new plants is expected to increase in the near future. These plants produce a waste concentrate stream in the vicinity of 38,000–57,000 m^3/day. Since these plants are placed far from the coast, direct discharge of the brine will easily affect the surrounding ecosystem. So, the issue is to develop energy efficient desalination minimizing discharge of brine and making environment friendly. More research is needed to develop cost effective and environment friendly desalination technology [68, 69].

7. Development opportunities

Pressures on water resources are increasing with the expanding scale of global development. Impacts from these pressures range from ecological and hydrological consequences of over-allocation of river basins and groundwater aquifers, to public health consequences and ecological damage arising from water quality deterioration [70].

The core concern is that demand for food and water is increasing across globe. Scarcity of future freshwater generation capacity and escalating costs of exploitation are great challenges. The problem would be further aggravated due to the effect of climate change and environmental impact. Thus, the fundamental policy and management concerns are how the available water resources could be managed more sustainably to enhance the efficiency of food production and to safeguard environmental systems and their provision of goods and services. In the face of the growing scarcity of water resources and the need for better management emphasis should be given on increasing current water use efficiency.

When considering economic efficiency of water resource use from a sustainability point of view as 'Scarce natural capital' it is important that water must be managed in such a way as not to reduce the opportunities for potential use by future generations. In this context, in addition to water use efficiency, it is much important to consider water withdrawal and use for irrigation purposes can have negative impacts on wetlands, aquatic ecosystems and corresponding ecological functions. Negative impacts also include external costs, such as those from water logging, salinity intrusion and soil erosion, which are also not usually incorporated into the economic price of irrigation water. Furthermore, even though water is being used more efficiently, the ecological limits to water use must be considered.

Water resources and effects are often non-marketed. It is much important to ensure that the 'true' economic values of such resources are accounted for making decisions on investment for water linked with environmental issue. Water productivity will have to be enhanced significantly in the coming decades via efficiency gains enabled through economic measures such as valuation, pricing and trading, as well as through technological innovation.

Unsustainable development pathways and governance failures have generated immense pressures on water resources, affecting its quality and availability, and in turn compromising its ability to generate social and economic benefits. The

Objective	Strategy	Actions
1.Balancing water supply and demand for sustainable inclusive growth	1. Water supply management	(i) Additional surface irrigation development from rivers improvement, (ii) Augmentation of surface water supply to Municipality of Dhaka, (iii) Deep ground water development, (iv) wastewater reuse
	2. Water demand management	(i) Adopt less water demanding crops, (ii) Irrigation water saving, technologies, (iii Domestic and industrial water saving
	3. Resource planning and protection	(i) River basin management, plans, (ii) Water resource development and protection in river basin
2. Maintaining water quality for health, livelihood and ecosystem	1. Enhancing fresh water flow	(i) Off take stabilization of the river Jamuna and flow augmentation in the north and central region, (ii) River management and smart dredging
	2. Pollution control and treatment	(i) Expansion of clean industrial production technologies, (ii) Low impact agriculture and aquaculture, (iii) Urban wastewater master plan, (iv) Monitoring permitting and control, (v) River basin environmental management plan

Table 8.
Development Strategies formulated in Bangladesh Delta Plan 2100.

planet's capacity to sustain the growing demands for freshwater is being challenged, and there can be no sustainable development unless the balance between demand and supply is restored and water quality is maintained for health, livelihoods and ecosystem which is addressed in the recent Fresh Water Development Strategies formulated in Bangladesh Delta Plan 2100 [23] (**Table 8**).

A number of options could be suggest for developing global water resources, enhancing water use efficiency and mitigate adverse impact of climate change on water availability and increasing agricultural productivity in the globe: (1) Augmentation of surface water through excavation of rivers, water bodies, development of water reservoirs, improved drainage, saline intrusion control, flood management and recharge of ground water. (2) Use water saving technology for improving efficiency of water and install facilities to reduce distribution losses in the crop field. Activities should include: (i) Reduce water losses in existing schemes through improved water management (capacity building of water management organizations), development of water saving techniques or rehabilitation of existing schemes.(3) Reduce impact of saline water intrusion in the main land and enhance river water flow. The focused activities are management of embankment and tidal river; expansion of surface water irrigation with construction of reservoir for monsoon water; and improved brackish water resource management practices. (4) Development of less water consuming and drought tolerant crop varieties, (5) Conservation of water resources for future use, (6) Wastewater treatment for reuse (7) Development and utilization of cost effective environment friendly desalination technology. (7) Development of climate smart and water precision agriculture and (7) Research on technology generation and dissemination.

8. Policy and institution

Favorable policy and institutional climate is needed for enhancing efficiency, conservation and sustainability of global water resource and increased food production in 2030 and 2050. The prospects for the implementation of sustainable water management policies to reverse degradation trends and conserve resources for the future will be effective if appropriate institutional set up could be established. Collaboration would be needed among the national and international water institutions and governments.

Institutional and political measures are further important building blocks for improved water management. In many regions, water is seriously under-valued, especially in the agricultural sector. This is one major reason for over-use and wastage. There is often a lack of well-defined property rights or are not implemented.

A range of technical and institutional solutions might be available to increase food production by almost 70% by 2050; to feed the increased population, reduce hunger and improve livelihoods for the poorest; and to minimize or mitigate degradation of land and water and of the broader ecosystems. They need to be adapted to local conditions and socio-economic contexts. Improved planning, linked to smart incentive packages, can then establish a framework for investment that assigns agreed values to natural capital. On this basis, land and water management that is efficient, equitable and sustainable can be encouraged at all levels. Some of the institutional development options could be suggested are: (1) Capacity development of farmers 'water associations, (2) Capacity development of municipal water distribution agency and water development institutions, (3) Adapting participatory water resource development planning for sustainable water management at local, regional and global. At all level, financing is required for increased levels of investment, and (4) Promoting Public Private Partnership in water sector.

Policies, institutions and implementation strategies should be adjustment at global, national and local levels to develop capacities of organizations and farmers with the knowledge and financial resources. Knowledge sharing at local, national and global levels focusing on land and water systems development will foster socio-economic growth across the globes reducing food insecurity and poverty.

9. Conclusions

World population will be 8.6 billion in 2030 and 9.8 billion in 2050. An additional 2.4 billion people projected to be added to the global population between 2015 and 2050 of which 1.3 billion will be added from Africa and 0.9 billion people from Asia. The growth rates of population and income are the important drivers of world food demand.

As income grows in developing world consumption of starchy-staple foods will decline while spending on animal-based products increases. It implies that income distribution matters for aggregate world food demand.

World cereal equivalent (CE) food demand is projected to be around 10,094 million tons in 2030 and 14,886 million tons in 2050. On the supply side, CE food production is projected to be 10,120 million tons in 2030 and 15,970 million tons in 2050. The world CE food demands would change by 134% from the base year of 2009 while CE food production would change by 151% and thus food production would grow slightly faster than demand yielding a positive gap of 7%. India is first leading country creating largest share of world food demand (24.3%) followed by China (16.7%). Although Bangladesh is small country but densely populated and stood 8th with a share of 2% of world food demand.

The growth rate of global demand for cereals declined continuously during 1969 to 2007 from 2% per annum to 1.3% and projected to fall further to 1.2% in 2030 and to 0.9% in 2050 while world cereal demand would have a significant rise from 940 million tons from the base year 2005/2007 to 3 billion metric tons by 2050. Almost all the increases in the consumption of cereals will come from the developing countries. The developing countries surpassed developed ones in total cereal production in the early 1990s: they now account for 56% of world production and the share will increase to 60% in 2050.

Agriculture is the largest water user accounting 70% of global water withdrawal and rest 30% is used by municipal, energy and industrial sectors. Global water demand is projected increase by 55% between 2000 and 2050 from 3500 to 5425 km^3. In addition to demand from the agricultural sector, large increases are predicted for industry (400%), energy production (140%) and domestic use (130%) Changing diet toward meat would enhance global water demand for growth of livestock sector and would cause scarcity of fresh water in many areas of the world. The potential for further expansion of irrigation is limited. There are plenty of renewable water resources globally; but they are extremely scarce in regions such as the Near East/North Africa, or Northern China, where they are most needed. Global water scarcity is growing more severe recent years. Research has demonstrated that two-thirds of the world's populations currently live in areas that experience water scarcity.

The availability of water resources is intrinsically linked to water quality. Evidence show that severe pathogen pollution affects around one third of all river stretches in Africa, Asia and Latin America. Release of agrochemicals, animal waste and anthropogenic activities are polluting fresh water, marine ecosystem and ground water.

Climate change will have adverse impact on world water resources through changing temperature, precipitation, melting snow, river flow, flood and drought. There are wide range of variability of these climatic events and vulnerability across various regions of the globe. Climate variability and extremes are negatively affecting agricultural productivity globally. With the existing technology it will be difficult to boost food production further in the future, specifically during 2030 and 2050. A technological breakthrough will be needed with introduction of climate resilient HYVs of wheat and rice to transform global agriculture to feed the increased population in the future.

Post-harvest loss in food is huge accounting around one-third of all production in developing countries. Food availability could be enhanced and made sustainable through reducing post-harvest loss with increased investments in market infrastructure, value addition, and food processing and promoting international trade.

Over one-third of the world's population lacks access to safe drinking water. Currently large number of desalination plants are operating worldwide and providing water for more than 300 million people. These desalination remains an energy intensive process and future costs will continue to depend on the price of both energy and desalination technology. These plants release daily huge brine whose disposal is costly and adversely affect surrounding eco-system.

With rising global water demand the policy and management concerns are manage water resource more sustainably to enhance the efficiency of food production and safeguard environmental systems. Emphasis should be given on increasing water use efficiency and conservation of water resources and ecology. A number of options are suggested for developing global water resources, enhancing water use efficiency and mitigate adverse impact of climate change on water availability in the globe and enhancing food production.

Author details

Sheikh Mohammad Fakhrul Islam* and Zahurul Karim
Modern Food Storage Facilities Project, World Bank, Dhaka, Bangladesh

*Address all correspondence to: smfakhruli@gmail.com

References

[1] Foresight. The Future of Food and Farming. Executive Summary, The Government Office for Science, London. 2011. Available from: https://www.gov.uk/government/publications/future-of-foodand-farming

[2] To H, Grafton RQ. Oil prices, biofuels production and food security: Past trends and future challenges. Food Security. 2015;**7**(2):323-336

[3] UNEP. Assessing global land use: Balancing consumption with sustainable supply. In: Bringezu S, Schütz H, Pengue W, O'Brien M, Garcia F, Sims R, et al., editors. A Report of the Working Group on Land and Soils of the International Resource Panel. Nairobi: United Nations Environment Programme; 2014

[4] Islam SMF, Talukder RK. Projections of food demand and supply in Bangladesh: Implications on food security and water demand. International Journal of Sustainable Agricultural Management and Informatics. 2017;**3**(2):125-153

[5] Grafton RQ, Williams J, Jiang QM. Food and water gaps to 2050: Preliminary results from the global food and water system (GFWS) platform. Food Security. 2015;**7**(2):209-220

[6] Zhou Y, Staatz J. Projected demand and supply for various foods in West Africa: In addition to demand from the agricultural sector, large increases are predicted for industry (400%), energy production (140%) and domestic use (130%) implications for investments and food policy. Food Policy. 2016;**61**(1):198-212

[7] Climate Change. Impacts, Adaptation, and Vulnerability, Working Group II Contribution. Fifth Assessment Report of the Intergovernmental Panel on Climate Change. Cambridge: Cambridge University Press; 2014

[8] United Nations, Department of Economic and Social Affairs, Population Division. World Population Prospects: The 2015 Revision. New York: United Nations; 2015. Available from: https://en.wikipedia.org/wiki/Projections_of_population_growth#cite_note-UN-WPP-2015-4

[9] World Population Prospects: The 2017 Revision: Key Findings and Advance Tables. New York: United Nations; 2017. Available at: https://esa.un.org/unpd/wpp/publications/files/wpp2017_keyfindings.pdf

[10] Godfray HC. Food for thought. Proceedings of the National Academy of Sciences. 2011;**108**(50):19845-19846

[11] Rask K, Rask N. Economic development and food production–consumption balance: A growing global challenge. Food Policy. 2011;**36**(2):186-196

[12] Cirera X, Masset E. Income distribution trends and future food demand. Philosophical Transactions of the Royal Society, B. 2010;**365**: 2821-2834. DOI: 10.1098/rstb.2010.0164

[13] Hertel T, Baldos U, van der Mensbrugghe D. Predicting long-term food demand, cropland use and prices. Annual Review of Resource Economics. 2016;**8**(1):417-441

[14] Lampe M et al. Why do global long-term scenarios for agriculture differ? An overview of the AgMIP global economic model intercomparison. Agricultural Economics. 2014;**45**(1):3-20

[15] Hertel T, Baldos U. Global Change and the Challenges of Sustainably Feeding a Growing Planet. Switzerland: Springer, International Publishing; 2016

[16] Gouel C, Guimbard H. Nutrition Transition and the Structure of Global

Food Demand. IFPRI Discussion Paper, 1631. Washington DC: International Food Policy Research Institute; 2017

[17] Emiko F, Will M. Economic Growth, Convergence, and World Food Demand and Supply, Policy Research Working Paper 8257, World Bank; 2017

[18] Alexandratos N, Bruinsma J. World Agriculture Towards 2030/2050, ESA Working Paper No. 12-03. Food and Agriculture Organization of the United Nations; 2012

[19] Amarasinghe UA, Sharma BR, Muthuwatta L, Khan ZH. Water for Food in Bangladesh: Outlook to 2030. Colombo, Sri Lanka: International Water Management Institute (IWMI); 2014. 32p. (IWMI Research Report 158

[20] The United Nations World Water Development Report. Wastewater: The Untapped Resource, UNESCO. 2017. Available from: https://unesdoc.unesco.org/ark:/48223/pf0000247153 [Accessed: 02 February 2019]

[21] The UN World Water Development Report 2015. Water for a Sustainable World, UNESCO; 2015

[22] OECD. Environmental Outlook to 2050. OECD Publishing; 2012 Available from: https://read.oecd-ilibrary.org/environment/oecd-environmental-outlook-to-2050_9789264122246-en#page1

[23] Bangladesh Delta Plan 2100. Government of the People's republic of Bangladesh. Dhaka: Bangladesh Planning Commission; 2017

[24] Integrated agricultural strategic plan for Teesta basin region in Bangladesh. Government of Bangladesh, IFPRI; 2016

[25] Mekonnen MM, Hoekstra AY. Four billion people facing severe water scarcity. Science Advances. 2016;**2**:e1500323

[26] A Snapshot of the World's Water Quality: Towards a Global Assessment. UNEP; 2016

[27] Kumar P, Masago Y, Mishra BK, Jalilov S, Emam AR, Kefi M, et al. Current assessment and future outlook for water resources considering climate change and a population burst: A case study of Ciliwung River, Jakarta City, Indonesia. Water. 2017;**9**(6):410. DOI: 10.3390/w9060410. Available from: https://www.mdpi.com/2073-4441/9/6/410/htm [Accessed: 02 February 2019]

[28] Climate Change 2014. Synthesis Report Fifth Assessment Report. Intergovernmental Panel on Climate Change; 2019

[29] Milly PCD, Wetherald RT, Delworth TL. Increasing risk of great floods in a changing climate. Nature. 2002;**415**:514-517

[30] Pall P, Aina T, Stone DA, Stott PA, Nozawa T, Hilberts AG, et al. Anthropogenic greenhouse gas contribution to flood risk in England and Wales in autumn 2000. Nature. 2011;**470**:382-385

[31] Carter JG, Cavan G, Connelly A, Guy S, Handley J, Kazmierczak A. Climate change and the city: Building capacity for urban adaptation. Progress in Planning. 2015;**95**:1-66

[32] IPCC. Climate Change 2013: The Physical Science Basis. Contribution of Working Group I to the Fifth Assessment Report of the Intergovernmental Panel on Climate Change. In: Stocker TF, Qin D, Plattner GK, Tignor M, Allen SK, Boschung J, Nauels A, Xia Y, Bex V, Midgley PM, editors. Cambridge, UK: Cambridge University Press; 2013

[33] Abbaspour KC, Faramarzi M, Ghasemi SS, Yang H. Assessing the impact of climate change on water resources in Iran. Water Resources Research. 2009;**45**:W10434

[34] Vicuna S, Dracup J. The evolution of climate change impact studies on hydrology and water resources in California. Climatic Change. 2007;**82**:327-350

[35] Liu Z, Xu Z, Huang J, Charles SP, Fu G. Impacts of climate change on hydrological processes in the headwater catchment of the Tarim River basin, China. Hydrological Processes. 2010;**24**:196-208

[36] Wang B, Zhang M, Wei J, Wang S, Li S, Ma Q, et al. Changes in extreme events of temperature and precipitation over Xinjiang, Northwest China, during 1960-2009. Quaternary International. 2013;**298**:141-151

[37] Climate Change 2013. The Physical Science Basis: Working Group I, Intergovernmental Panel on Climate Change–Business & Economics; 2014

[38] Luo M, Meng F, Liu T, Duan Y, Frankl A, Kurban A, et al. Multi–model ensemble approaches to assessment of effects of local climate change on water resources of the Hotan River Basin in Xinjiang, China. Water. 2017;**9**:584. DOI: 10.3390/w9080584. Available from: https://www.mdpi.com/2073-4441/9/8/584 [Accessed: 02 February 2019]

[39] Lee SH, Yoo SH , Choi JY, Bae S. Assessment of the impact of climate change on drought characteristics in the Hwanghae plain, North Korea using time series SPI and SPEI: 1981-2100, Water 9(8):579 · August 2017. DOI: 10.3390/w9080579

[40] Climate Change. The Physical Science Basis: Working Group I Contribution to the Fourth Assessment Report of the IPCC, Intergovernmental Panel on Climate Change. Cambridge: Cambridge University Press; 2007

[41] IPCC. Third Assessment Report on Climate Change. Climate change, Working Group II: Impacts, Adaptation and Vulnerability, Intergovernmental Panel on. Cambridge University Press; 2001

[42] Klein Tank AMG, Wijngaard J, van Engelen A. Climate of Europe: Assessment of observed daily temperature and precipitation extremes. European Climate Assessment (ECA). 2002

[43] United Nations. Economic Commission for Africa 2006. African Water Development Report 2006. Final Version. http://hdl.handle.net/10855/22091 [Accessed: 14 November 2009]

[44] Urama KC, Ozor N. Impacts of climate change on water resources in Africa: The role of adaptation. African Technology Policy Studies Network (ATPS). 2010

[45] Yates DN. Climate change impacts on the hydrologic resources of South America: An annual, continental scale assessment. Climate Research. 1997;**9**:147-155

[46] Karmalkar AV, Bradley RS, Diaz HF. Climate change in Central America and Mexico: Regional climate model validation and climate change projections. Climate Dynamics. 2011;**37**(3-4):605-629

[47] World Bank. Environmental Health in Nicaragua. Addressing Key Environmental Challenges. Washington, D.C: The World Bank; 2013

[48] Luce CH, Lopez-Burgos V, Holden Z. Sensitivity of snowpack storage to precipitation and temperature using spatial and temporal analog models. Water Resources Research. AGU Publication; 2014;**50**:9447-9462. DOI: 10.1002/2013WR014844

[49] Cliftona CF, Dayb KT, Lucec CH, Grantd GE, Safeeqe M, Halofskyf JE,

Staaba BP. Effects of climate change on hydrology and water resources in the Blue Mountains, Oregon, USA

[50] Adams HD, Luce CH, Breshears DD, et al. Ecohydrological consequences of drought- and infestation-triggered tree die-off: Insights and hypotheses. Ecohydrology. 2012;**5**:145-159

[51] Vose JM, Clark JS, Luce CH. Introduction to drought and US forests: Impacts and potential management responses. Forest Ecology and Management. 2016;**380**:296-298. DOI: 10.1016/j.foreco.2016.09.030

[52] Jones RN, Chiew FH, Boughton WC, Zhang L. Estimating the sensitivity of mean annual runoff to climate change using selected hydrological models. Advances in Water Resources. 2006;**29**(10):1419-1429. http://dx.doi.org/10.1016/j.advwatres.2005.11.001

[53] FAO. The Future of Food and Agriculture—Trends and Challenges. Rome; 2017

[54] FAO, IFAD, UNICEF, WFP and WHO. The State of Food Security and Nutrition in the World 2018. Building Climate Resilience for Food Security and Nutrition. Rome: FAO; 2018

[55] Daryanto S, Wang L, Jacinthe P-A. Global synthesis of drought effects on maize and wheat production. PLoS One. 2016;**11**(5):e0156362. Available from: https://www.ncbi.nlm.nih.gov/pmc/articles/PMC4880198/

[56] Hansen J, Ruedy R, Sato M, Lo K. Global surface temperature change. Reviews of Geophysics. 2010;**48**:RG4004

[57] Nelson GC, Rosegrant MW, Koo J, Robertson R, Sulser T, Zhu T, et al. Climate Change: Impact on Agriculture and Costs of Adaptation. Vol. 21. Washington, DC, USA: IFPRI; 2009

[58] Li M. Climate Change to Adversely Impact Grain Production in China by 2030. Vol. 2018. Washington, DC, USA: IFPRI; 2018

[59] Karim Z, Hussain SG, Ahmed M. Assessing impacts of climatic variations on food grain production in Bangladesh. Water, Air and Soil Pollution, Kluwer Academic Publishers. 1996;**92**:53-62

[60] Li X, Takahashi T, Suzuki N, Kaiser HM. Impact of climate change on maize production in Northeast and Southwest China and risk mitigation strategies. APCBEE Procedia. 2014;**8**:11-20

[61] Leng G, Huang M. Crop yield response to climate change varies with crop spatial distribution pattern. Scientific Reports. 2017;**7**:1463

[62] Lobell DB, Burke MB. On the use of statistical models to predict crop yield responses to climate change. Agricultural and Forest Meteorology. 2010;**150**:1443-1452

[63] Bassu S, Brisson N, Durand JL, Boote K, Lizaso J, Jones JW, et al. How do various maize crop models vary in their responses to climate change factors? Global Change Biology. 2014;**20**:2301-2320

[64] Gammans M, Mérel P, Ortiz-Bobea A. Negative impacts of climate change on cereal yields: Statistical evidence from France. Environmental Research Letters. 2017;**12**:054007

[65] Tao F, Zhang Z, Xiao D, Zhang S, Rötter RP, Shi W, et al. Responses of wheat growth and yield to climate change in different climate zones of China, 1981-2009. Agricultural and Forest Meteorology. 2014;**189**:91-104

[66] Schwarzenbach RP, Escher BI, Fenner K, Hofstetter TB, Johnson CA, Von Gunten U, et al. The challenge of micropollutants in aquatic systems. Science. 2006;**313**:1072-1077

[67] Greenlee LF, Lawler DF, Freeman BD, Marrot B, Moulin P. Reverse osmosis desalination: Water sources, technology, and today's challenges. Water Research. 2009;**43**:2317-2348

[68] Qiu TY, Davies PA. The scope to improve the efficiency of solar-powered reverse osmosis. Desalination and Water Treatment. 2011;**35**:14-32

[69] Elimelech M, Phillip WA. The future of seawater desalination: Energy, technology, and the environment. Science. 2011;**333**:712-717

[70] Falkenmark M, Rockström J. Balancing Water for Man and Nature: The New Approach to Ecohydrology. U.K: EarthScan; 2004

Chapter 5

A Novel Low-Temperature Thermal Desalination Technology Using Direct-Contact Spray Method

Qian Chen, Muhammad Burhan, Muhammad Wakil Shahzard, Raid Alrowais, Doskhan Ybyraiymkul, Faheem Hassan Akhtar, Yong Li and Kim Choon Ng

Abstract

Due to the emerging water crisis, the global desalination capacity has been expanding exponentially in the past few decades, leading to substantial amount of primary energy consumption. Therefore, the exploration of energy-efficient desalination processes and alternative energy sources has been the subject of great research interests. The spray-assisted low-temperature desalination (SLTD) system is a novel method for desalination that enables efficient renewable energy utilization. It works on the direct-contact spray evaporation/condensation mechanism and uses only hollow chambers. The merits include enhanced heat and mass transfer, lower initial and operational costs, and reduced scaling and fouling issues. This chapter presents a study on the SLTD system driven by sensible heat sources. The working principle of the system will be introduced first. Then a thermodynamic analysis will be presented to obtain the freshwater productivity under different design and operational conditions. Additionally, the energy utilization level will be quantified to highlight the energy wastage when operating with sensible heat sources. Afterward, the system configuration will be modified to maximize the utilization of sensible heat sources and promote productivity. Finally economic viability of the modified design will be evaluated.

Keywords: direct-contact spray, thermal desalination, sensible heat source, thermodynamic analysis, internal heat recovery

1. Introduction

Freshwater is the key resource for the continuation of human society. With the growth of world population, the world water consumption has been increasing exponentially in the past decades [1, 2]. Meanwhile, freshwater resources on the earth are limited, and they are degrading and depleting due to overexploration and environmental pollution [3]. Consequently, the global water deficit is becoming more and more severe. By 2030, the world water deficit is expected to reach 2700 billion m^3/year [4], and the population that will suffer from water shortage will exceed 1.6 billion [5]. Therefore, it is of ultimate importance to develop new and sustainable sources for freshwater supply.

Seawater desalination is one of the most promising solutions to the issues associated with water shortage [6]. It is the process that separates a portion of freshwater from the seawater with the input of energy or work. The great potential of seawater desalination lies in the wide availability of seawater. More than 70% of the earth surface is covered by oceans. More importantly, most of the cities that are facing water shortage are located along the coast [7], and nearly 40% of the world population lives near the sea [8]. With the advances in desalination technologies, the great potential of desalination has been translating into an expanding global desalination capacity. So far, more than 16,000 desalination plants have been installed in nearly 150 countries, and the overall desalination capacity has exceeded 90 million m^3/day [9].

Existing desalination technologies in the market can be divided into two categories, namely, membrane-based processes and thermally driven processes. The membrane-based process, represented by reverse osmosis (RO), uses a semipermeable membrane to separate freshwater from seawater. The membrane only allows water molecules to pass, leaving behind the salt. In a RO system, the seawater is pressurized to overcome the osmotic pressure and drive the diffusion of water molecules. RO systems not only exhibit low-energy consumption but also have small plant footprint, making it the dominating technology in the desalination market. By 2016, the market share of RO has exceeded to 63% [9].

Thermally driven processes, as presented by multi-effect distillation (MED) and multistage flash distillation (MSF), separate water and salt by evaporation and subsequent condensation. Since salts are not volatile, thermally driven processes are able to reject almost 100% of the dissolved salt and achieve a very high distillate quality. The energy consumption of MED and MSF is much higher than RO due to the high latent heat of vaporization. However, thermally driven processes utilize low-grade heat instead of electricity. With various waste heat available in different industrial processes, thermal desalination processes are sometimes more appealing than RO. Moreover, thermal processes are less sensitive to the change of feed salinity, and they are able to handle harsh feed conditions where RO is not applicable.

One major barrier that hinders the wider application of MED and MSF is the high initial plant cost, which limits them to large-scale operations. However, cost-effective low-grade heat sources, such as industrial waste heat, are often available in a small amount. Therefore, it is of great impetus to develop small-scale thermally driven processes. Humidification-dehumidification (HDH) processes and membrane distillation (MD) are two emerging technologies that are suitable for small-scale operation. But both processes are facing key challenges that should be overcome before wider application is possible. The productivity of HDH is limited by the small vapor-carrying capability of air, and the footprint size is relatively large due to small heat and mass coefficients between wet air and condenser surface [10]. On the other hand, MD are facing the issues of small distillate flux, membrane degradation due to scaling and fouling, and relatively low thermal efficiency.

While on-going research efforts are being conducted to address the issues faced by HDH and MD, the development of more advanced thermal processes is also of importance. The spray-assisted low-temperature desalination (SLTD) technology is a recently proposed method that mitigates the issues faced by conventional thermal processes. It employs direct-contact evaporation/condensation method, thus eliminating metallic surfaces inside the system. The merits include (1) promoted heat and mass transfer due to direct contact between water and vapor, (2) reduced scaling and fouling potential, and (3) lower initial and operational costs. In our previous publications, the technical viability of the direct spray method has been demonstrated experimentally [11, 12] and analytically [10, 13]. The thermodynamic performance [14–17] and economic viability [16] have also been evaluated. However, all of these studies employ steam as the heat source, while none of them have ever considered sensible heat sources.

This chapter specially investigates an SLTD system driven by sensible heat sources. Different from steam, sensible heat sources suffer from a temperature drop in order to release energy. To sustain a temperature difference for heat transfer, the heat source is always at a higher temperature than the medium that is being heated. Consequently, a portion of enthalpy in the heat source is usually left unused. Therefore, conventional energy efficiency measurements are not applicable for sensible heat sources since they only look into heat extraction from the heat sources and neglect un-extracted energy (which is negligible in steam-driven systems). In the following sections, the productivity of the SLTD system will be evaluated under different design and operational conditions. The energy utilization level will also be calculated to quantify the amount of unused enthalpy in the heat source. Then a modified configuration will be proposed to enhance energy utilization and boost freshwater production. A cost analysis will also be conducted to evaluate the economic viability of the proposed configuration.

2. System description and mathematical modeling

Figure 1 shows a simplified schematic of the proposed system, consisting of a series of evaporator-condenser stages, three sets of heat exchangers, and a vacuum

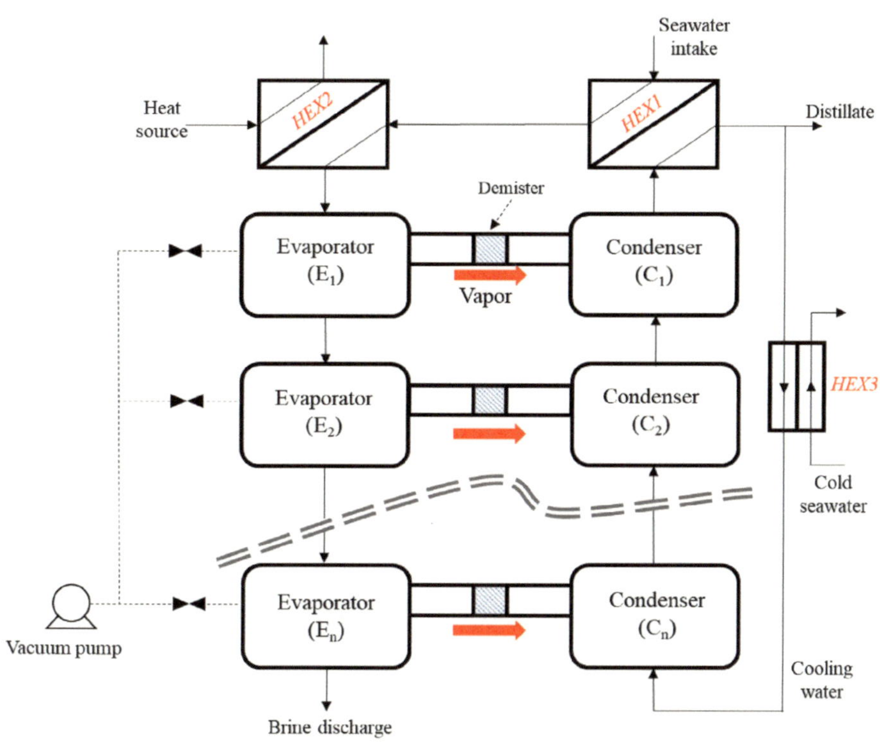

Figure 1.
Schematic of the spray-assisted low-temperature desalination system.

pump. Both evaporators and condensers are empty vessels operating under vacuum conditions. During operation, seawater is preheated externally and then sprayed into the evaporators, while cold cooling water is sprayed into the adjacent condensers. Driven by the partial vapor pressure difference, a portion of water evaporates from the seawater surface, travels to the condenser, and is condensed by the cooling water. The unevaporated seawater is then sprayed into the following evaporator, while the cooling water enters the previous condenser. The production stages are subjected to sequentially lowered pressure conditions so that the evaporation/condensation cycle is repeated. Finally the brine is disposed in the last evaporator, while the mixture of cooling water and distillate leaves from the first effect. Due to the accumulation of condensation heat, the mixed stream has a high temperature and is allowed to exchange heat with the intake seawater in *HEX1* to recover the condensation heat. Then the distillate is separated, and the cooling water is further cooled down in *HEX3* using the seawater before returning back to the stages. Meanwhile, the preheated seawater is directed to *HEX2* to be further heated to the desired temperature using an external heat source. The vacuum pump creates an initial vacuum condition at the beginning and removes the non-condensable gases dissolved in the seawater during operation.

The performance of the system can be predicted by analyzing heat and mass transfer between water and vapor in the vacuum environment as well as heat and mass balances among different components. **Figure 2** shows the schematic of the system components. The symbols that are used in the mathematical model are also included in **Figure 2**, while the governing equations are summarized in **Table 1**.

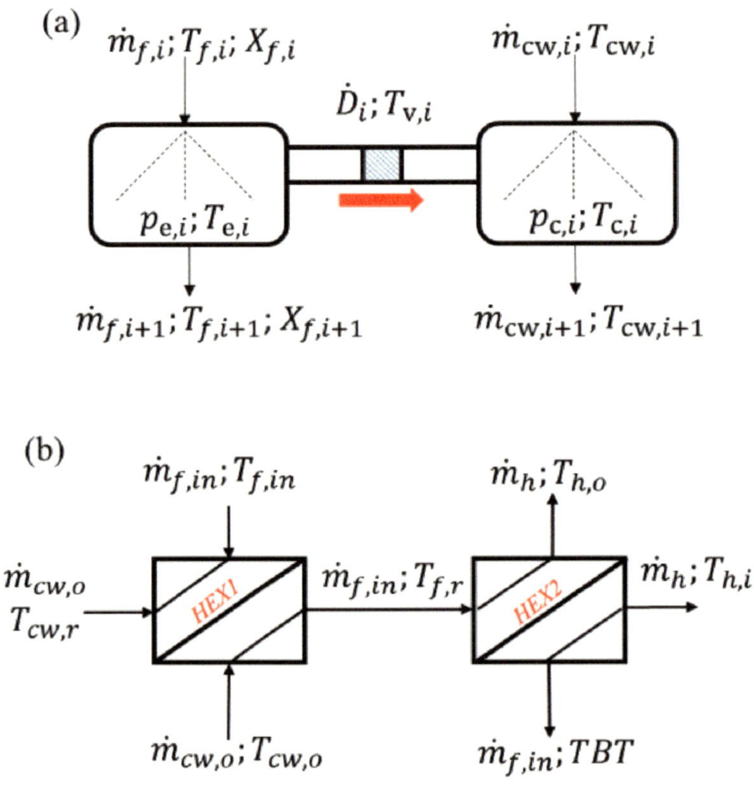

Figure 2.
Schematic of the system components with the main parameters: (a) production stages, and (b) heat exchangers.

Component	Equation	No.	Remarks
Evaporators	$T_{el,i+1} = T_{ev,i} + (T_{el,i} - T_{ev,i})\theta_{c,i}$	(1)	Seawater temperature change in evaporator; θ represents completeness of evaporation process [13]
	$T_{el,1} = TBT$	(2)	Inlet condition at the first evaporator
	$\dot{D}_{v,i} = \dot{m}_{el,i}\frac{c_{pel,i}(T_{el,i}-T_{el,i+1})}{h_{fge,i}}$	(3)	Amount of vapor produced in evaporator
	$\dot{m}_{el,i+1} = \dot{m}_{el,i} - \dot{D}_{v,i}$	(4)	Seawater mass conservation
	$\dot{m}_{el,i+1}X_{l,i+1} = \dot{m}_{el,i}X_{l,i}$	(5)	Salt mass conservation
Condensers	$T_{cl,i} = T_{cv,i} - (T_{cv,i} - T_{cl,i+1})\theta_{c,i}$	(6)	Cooling water temperature; θ represents the completeness of condensation process
	$T_{cl,n+1} = T_{cw,in}$	(7)	Inlet condition at last condenser
	$T_{cv,i} = T_{ev,i} - BPE_I - T_{loss}$	(8)	Temperature drop of vapor due to (1) boiling point elevation caused by dissolved salt and (2) temperature drop across the demister due to pressure drop
	$\dot{D}_{v,i} = \dot{m}_{cl,i}\frac{c_{pcl,i}(T_{cl,i}-T_{cl,i+1})}{h_{fgc,i}}$	(9)	Amount of vapor condensed on cooling water surface
	$\dot{m}_{cl,i} = \dot{m}_{cl,i+1} + \dot{D}_{v,i}$	(10)	Cooling water mass conservation
Heat exchangers	$\dot{m}_{sw,in}c_{p,sw}(T_{sw,r} - T_{sw,in}) = \dot{m}_{cw,o}c_{p,cw}(T_{cl,o} - T_{cl,r})$	(11)	Heat balance in *HEX1*
	$\dot{m}_{sw,in}c_{p,sw}(TBT - T_{sw,r}) = \dot{m}_h c_{p,h}(T_{h,i} - T_{h,o})$	(12)	Heat balance in *HEX2*

Table 1.
Model equations for the spray-assisted low-temperature desalination system.

3. Performance analysis

Employing the developed model, a 10-stage system operating at a top brine temperature of 70°C is firstly analyzed. Without loss of generality, the flowrate of the seawater is considered to be 10 kg/s. The flowrate of the cooling water and the heat source (considered to be hot water in this study) is equal to the feed flowrate in order to achieve the optimal system performance [14, 15]. The intake seawater is assumed to have a temperature of 25°C, and it will cool down the cooling water to 30°C in the counterflow heat exchanger (*HEX3*). **Figure 3** shows the temperatures for seawater and cooling water as well as freshwater productivity in each effect. It is obvious that seawater temperature drops successively along the stages and finally the brine is disposed at 33°C. On the other hand, the cooling water temperature increases in the reverse direction after absorbing the condensation heat and approaches the top brine temperature when leaving the first stage. The productivity also shows a descending trend, which is attributed to the drop of seawater flowrate after partial evaporation. The overall productivity is calculated to be 0.61 kg/s.

The productivity is expected to change when varying the number of stages and the top brine temperature. The former will change the efficiency of energy utilization, while the latter determines the total amount of energy that is available for

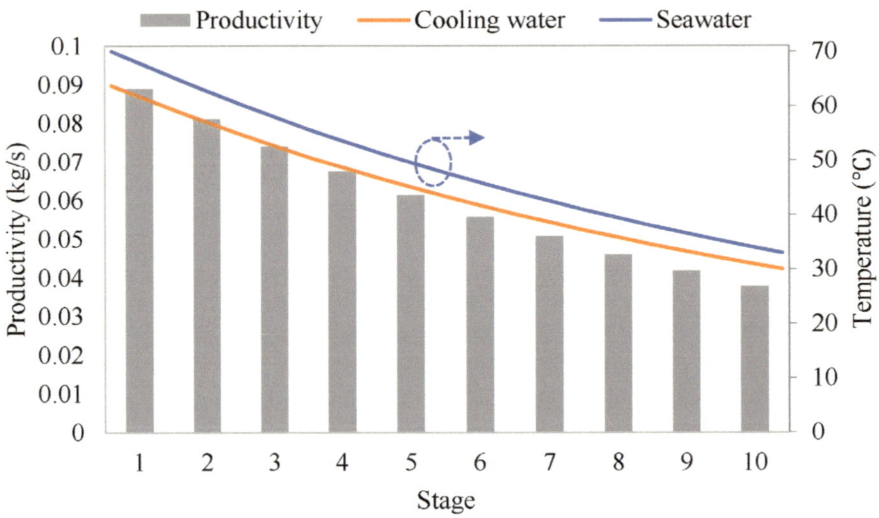

Figure 3.
Temperature and production profile for a typical system with 10 production stages.

evaporation. **Figure 4** shows the productivity under different conditions. The productivity firstly increases when the system has more operating stages, and the increasing trend gradually gets saturated. The reason is that increasing the number of stages improves only the energy efficiency, while the total amount of available heat is fixed. On the other hand, a higher top brine temperature will lead to remarkable improvement in productivity due to the availability of more heat source.

The majority of heating requirement is satisfied by the recovered condensation heat in *HEX2*, and the external heat source only undertakes a small portion. Take the 10-stage system operating at a *TBT* of 70°C as an example. As shown in **Figure 3**, the cooling water leaving the first condenser has a temperature of 63.7°C and is able to heat up the intake seawater to 58.7°C, considering an approach

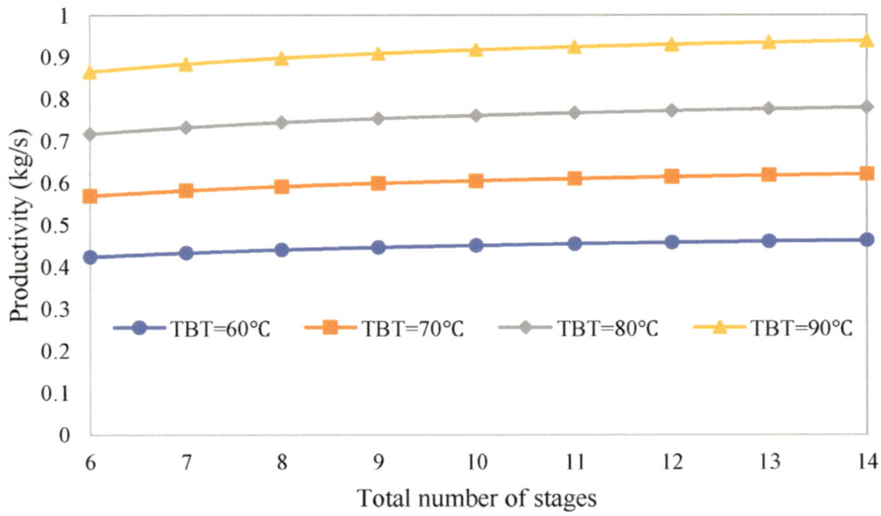

Figure 4.
Productivity under different top brine temperatures and operating stages.

temperature difference of 5°C in the heat exchanger. Therefore, the heat source only needs to heat up the seawater from 58.7 to 70°C, and its temperature drops from 75 to 63.7°C. Under such situation, only a very small portion of enthalpy is utilized from the heat source, while the remaining is left unused.

To quantify the efficiency of energy utilization, the level energy extraction from the heat source is calculated using the following equation:

$$Energy\ utilization = \frac{\dot{Q}_{utilized}}{\dot{Q}_{available}} = \frac{\dot{m}_h c_{p,h}(T_{h,i} - T_{h,o})}{\dot{m}_h c_{p,h}(T_{h,i} - T_{amb})} \tag{13}$$

The values of energy utilization under different design and operational conditions are plotted in **Figure 5**. Under all of the conditions, <20% of enthalpy is extracted from the heat source, while >80% is left unused. Such a low level of energy utilization results in significant wastage of the heat source and requires further optimization of the system.

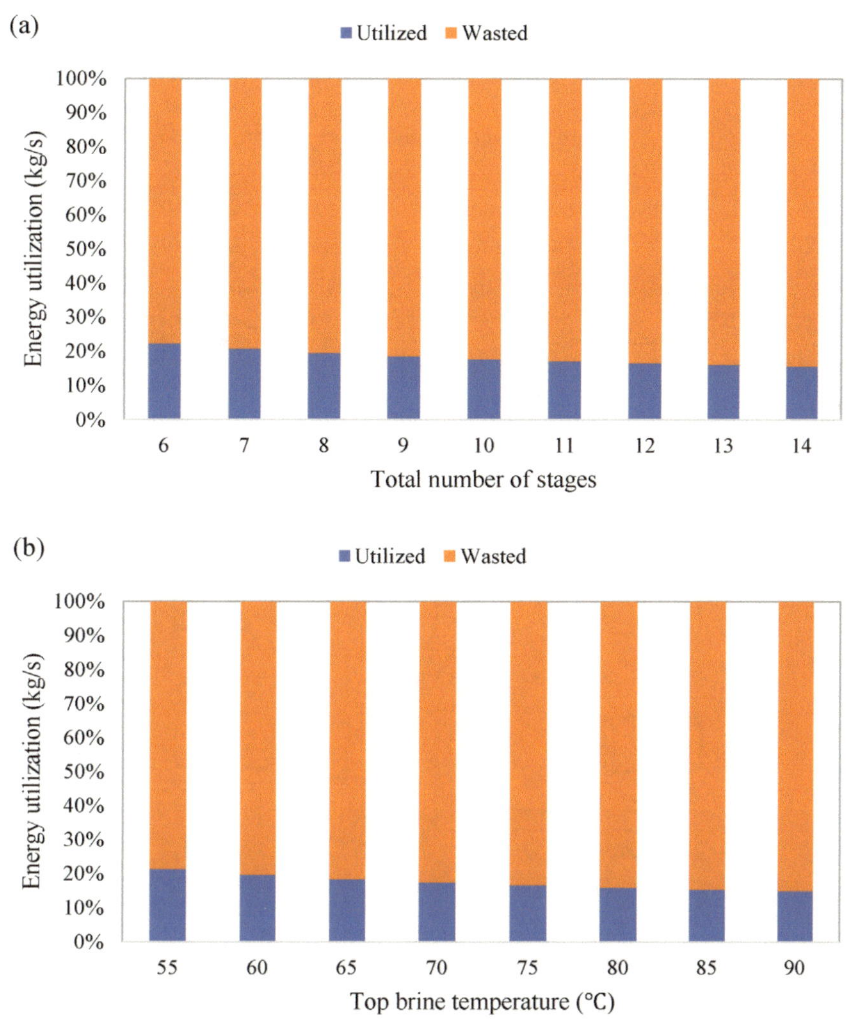

Figure 5.
Level of energy utilization under different (a) numbers of operating stages and (b) top brine temperatures.

4. System optimization

From the temperature profiles shown in **Figure 3**, it can be found that the cooling water in the earlier stages has a high temperature than the seawater in the later stages. For example, the cooling water leaving the second condenser has a temperature of 58.9°C, while the seawater enters the fifth evaporator at 51.4°C. If these two streams could exchange heat with each other, both the first and the fifth stages will have larger operating temperature differences, leading to a promoted productivity.

To facilitate such heat recovery, additional heat exchangers are added to the system, as demonstrated in **Figure 6**. For the first few condensers, the cooling water leaving the *i*th condenser is employed to preheat the seawater that is to be supplied into the *(i+NR)*th evaporator. To conserve heat exchanger area, a minimum temperature difference is set between the seawater and the cooling water. If the temperature difference between the two streams is below this value, no heat exchanger will be added for this evaporator-condenser pair.

The temperature variation for the seawater and cooling water, which are previously described by Eqs. (1) and (6), is now expressed as

$$T_{el,i+1} = \begin{cases} T_{ev,i} + (T_{el,i} - T_{ev,i})\theta_{e,i} & (i < NR) \\ T_{ev,i} + (T_{el,i} - T_{ev,i})\theta_{e,i} + \dfrac{\dot{Q}_{i-NR}}{\dot{m}_{el,i} c_p} & (i \geq NR) \end{cases} \tag{14}$$

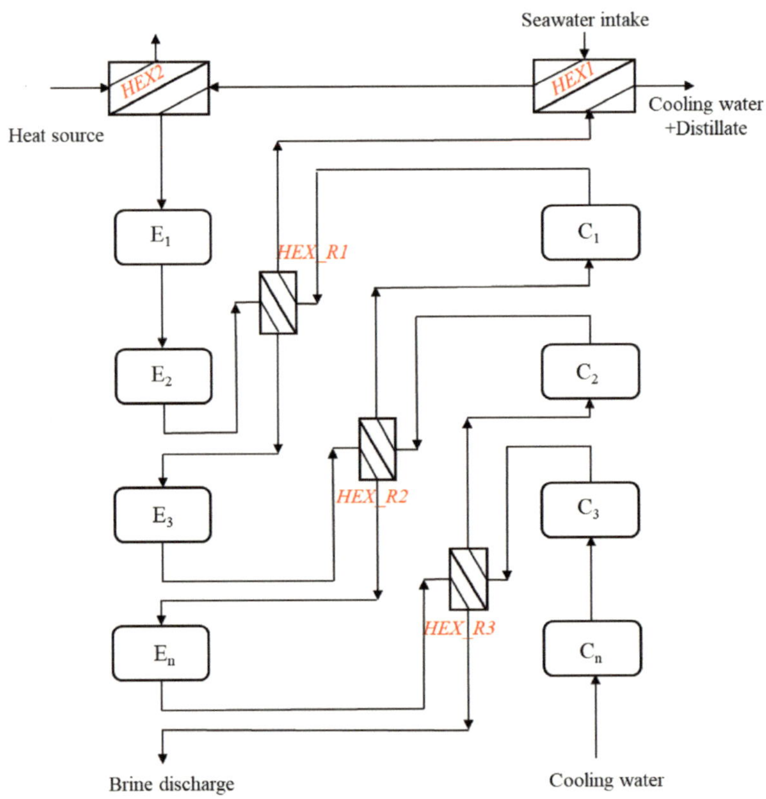

Figure 6.
Modified system with additional heat exchangers to enable internal heat recovery.

$$T_{cl,i} = \begin{cases} T_{cv,i} - (T_{cv,i} - T_{cl,i+1})\theta_{c,i} & (i > N_{total} - NR) \\ T_{cv,i} - (T_{cv,i} - T_{cl,i+1})\theta_{c,i} - \dfrac{\dot{Q}_i}{\dot{m}_{cl,i}c_p} & (i \le N_{total} - NR) \end{cases} \quad (15)$$

$$\dot{Q}_i = \begin{cases} \dot{m}_{cl,i}c_p(T_{cl,i} - T_{el,i+NR-1} - \Delta T_{HR})\theta_{c,i} & (T_{cl,i} > T_{el,i+NR-1} + \Delta T_{HR}) \\ 0 & (T_{cl,i} \le T_{el,i+NR-1} + \Delta T_{HR}) \end{cases} \quad (16)$$

Figure 7 shows the temperature profile for the modified system with 10 stages and a *TBT* of 70°C. The optimal value of *NR* is found to be 3 for this configuration. It is clearly shown in **Figure 7** that the feed seawater entering the 4th, 7th, and 10th evaporators are preheated by the cooling water leaving the 1st, 4th, and 7th condenser, respectively. As a result, productivity in the 4th, 7th, and 10th stages is significantly improved, as compared to the original system whose productivity shows a descending trend along the stages. The overall productivity has also been increased due to the elevated temperature gradient, from 0.61 kg/s to 0.74 kg/s.

Figure 8 presents the improvement of productivity for the modified system under different design and operational conditions. **Figure 8(a)** shows the performance under different numbers of operating stages. The *TBT* is kept fixed at 70°C, and the values of *NR* are varied according to the number of operating stages to achieve maximal fresh water yield. The modified system overperforms the original one with an increase ranging from 20 to 45%. Another difference is that the productivity does not increase monotonously with the number of operating stages for the modified system. This is attributed to the discontinuous values of optimal *NR* under different numbers of operating stages [18]. **Figure 8(b)** presents the system performance under different *TBT* for a 10-stage system. The productivity is improved by 55–78% compared to the original configuration. The higher the *TBT*, the more pronounced the improvement is.

As a result of the heat recovery action, the cooling water temperature leaving the first condenser is much lower. Consequently, less recovered condensation heat is available for seawater preheating in *HEX2*, and more heat input is required from the heat source. Therefore, the outlet temperature of the heating medium will be lower,

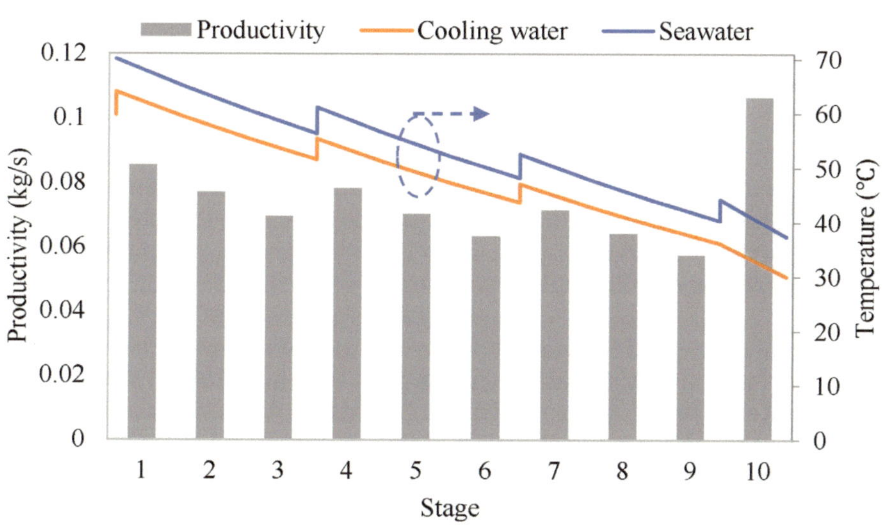

Figure 7.
Temperature and productivity profile for the modified system.

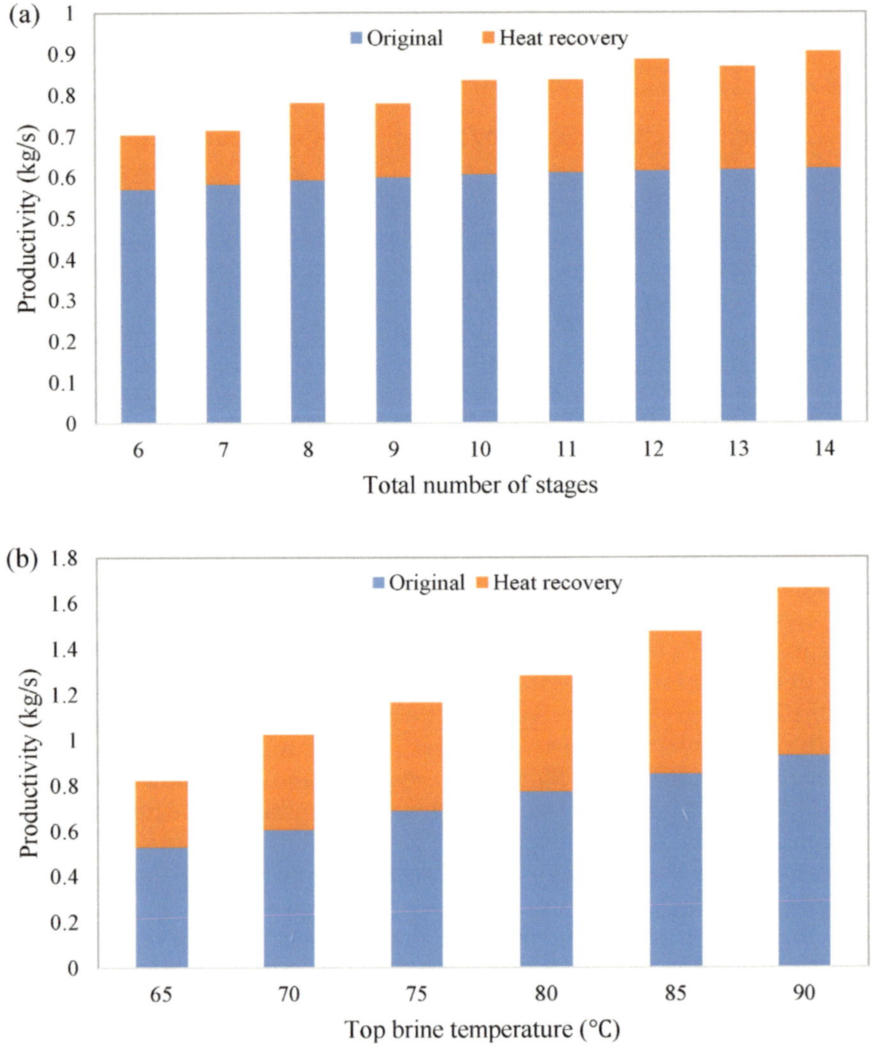

Figure 8.
Productivity improvement for the modified system under different (a) numbers of operating stages and (b) top brine temperatures.

indicating more heat extraction. Since the remaining energy cannot be further used, such change will not lead to additional energy cost and is acceptable. **Figure 9** compares the level of energy utilization before and after the heat internal heat recovery is conducted. It is clearly shown in **Figure 9** that the level of energy utilization is promoted by more than twice under all of the operating temperatures considered.

To evaluate the economic viability of the modified system, the costs for the system components are calculated for the original and the modified systems, as shown in **Table 2**. The calculation considers a 10-stage system operating at a *TBT* of 70°C. Compared with the original system, additional heat exchangers are added for internal heat recovery. Also, the pumping capacity will be higher to overcome the pressure drop across the newly added heat exchangers. As a result, the initial plant cost is increased by 9%. With an overall production increase of 23%, the unit cost for desalinated water will be lowered by 11%, making the modified configuration economic viable.

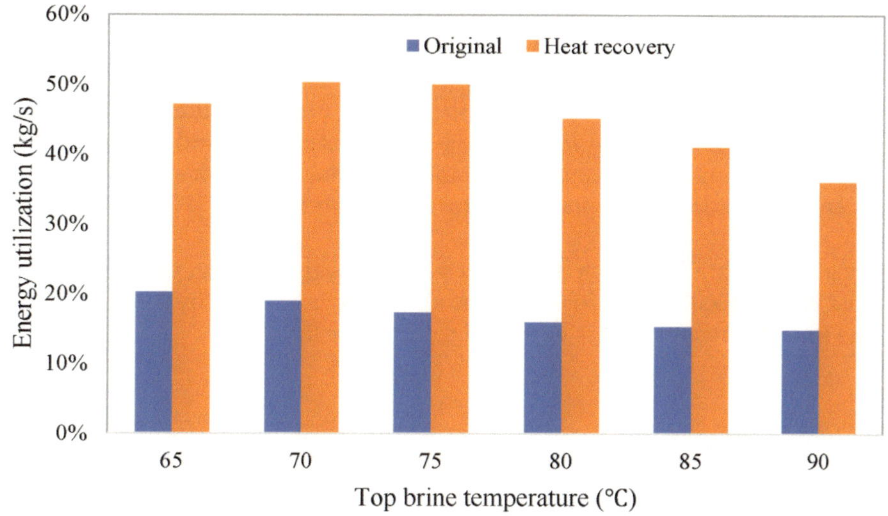

Figure 9.
Comparison of energy utilization for the heat source before and after heat recovery.

Component	Original	Modified	Difference
Evaporators ($)	32,785	32,785	–
Condensers ($)	9900	9900	–
Pumps ($)	3553	3560	0.2%
Heat exchangers ($)	23,574	30,065	28%
Total cost ($)	69,811	76,309	9%
Productivity (kg/s)	0.61	0.74	23%
Unit cost ($/(kgs^{-1})	115,295	102,621	-11%

Table 2.
Costs of system components before and after modification.

5. Conclusions

A spray-assisted low-temperature desalination system-driven source has been proposed and evaluated. The system employs hollow chambers for evaporation and condensation, which not only promotes heat and mass transfer but also reduces the plant cost. The system is analyzed, employing sensible heat as the energy source. The key findings of this study are summarized as follows:

1. Both the water temperatures and the productivity decrease monotonously along the stages in the original SLTD configuration. A higher cooling water outlet temperature results in a high level of heat recovery and reduces the heat input requirement.

2. The freshwater productivity is proportional to the top brine temperature due to more heat source available. More numbers of operating stages also boost productivity because of the promoted heat utilization. However, less than 20%

of enthalpy is extracted and utilized from the heat source, and the majority is left unused.

3. By using the cooling water to heat up the seawater in the intermediate stages, the temperature gradient in each stage becomes greater, and the productivity is promoted under different design and operating conditions. The level of energy utilization is also increased by more than twice.

4. By adding extra heat exchangers for internal heat recovery, the initial plant cost is increased by 9%, while the productivity is boosted by 23%. As a result, the unit freshwater cost is expected to decrease by 11%.

Acknowledgements

The authors gratefully acknowledge the generous funding from the (1) Water Desalination and Reuse Center (WDRC), King Abdullah University of Science and Technology (KAUST); (2) the National Research Foundation Singapore under its Campus for Research Excellence and Technological Enterprise (CREATE) program; and (3) the China Scholarship Council (CSC).

Conflict of interest

The authors declare no conflict of interest.

Nomenclature

BPE	boiling point elevation, °C
c_p	specific heat, J/kg
$\dot{D}$	production rate, kg/s
h_{fg}	latent heat of evaporation/condensation, J/kg
$\dot{m}$	mass flowrate, kg/s
N	total number of stages
NR	stage difference in heat recovery
Q	heat flux, W
X	salinity, kg/kg
T	temperature, °C
TBT	top brine temperature, °C

Greek letters

Δ	difference
θ	evaporation/condensation completion level

Subscripts

c	condenser/condensation
cl	cooling water
cv	vapor phase in condenser
cw	cooling water

e	evaporator
el	seawater in evaporator
ev	vapor in evaporator
h	heat source/hot water
HR	heat recovery
i	stage number
in	inlet
l	liquid phase
loss	temperature drop across the demister
r	heat recovery in *HEX1*
sw	feed seawater
v	vapor

Author details

Qian Chen[1*], Muhammad Burhan[1], Muhammad Wakil Shahzard[2], Raid Alrowais[1], Doskhan Ybyraiymkul[1], Faheem Hassan Akhtar[1], Yong Li[3] and Kim Choon Ng[1]

1 Water Desalination and Reuse Center, King Abdullah University of Science and Technology, Thuwal, Saudi Arabia

2 Northumbria University, Newcastle upon Tyne, United Kingdom

3 Institute of Refrigeration and Cryogenics, Shanghai Jiao Tong University, Shanghai, China

*Address all correspondence to: chen_qian@u.nus.edu

References

[1] Eltawil MA, Zhengming Z, Yuan L. A review of renewable energy technologies integrated with desalination systems. Renewable and Sustainable Energy Reviews. 2009; **13**(9):2245-2262

[2] Water U. Water in a changing world. In: The United Nations World Water Development Report 3. World Water Assessment Programme. Paris, France. 2009. p. 429

[3] Khawaji AD, Kutubkhanah IK, Wie J-M. Advances in seawater desalination technologies. Desalination. 2008;**221**(1-3):47-69

[4] Shahzad MW et al. Multi effect desalination and adsorption desalination (MEDAD): A hybrid desalination method. Applied Thermal Engineering. 2014;**72**(2):289-297

[5] Ferroukhi R et al. Renewable Energy in the Water, Energy & Food Nexus. Abu Dhabi: IRENA; 2015

[6] Shannon MA et al. Science and technology for water purification in the coming decades. In: Nanoscience and Technology: A Collection of Reviews from Nature Journals. World Scientific; 2010. pp. 337-346

[7] Ghaffour N. The challenge of capacity-building strategies and perspectives for desalination for sustainable water use in MENA. Desalination and Water Treatment. 2009;**5**(1-3):48-53

[8] Quteishat K. Desalination and water affordability. In: SITeau International Conference, Casablanca, Morocco. 2009

[9] Global Water Intelligence. IDA Desalination Yearbook 2017–2018. Global Water Intelligence; 2017

[10] Chen Q et al. Evaluation of a solar-powered spray-assisted low-temperature desalination technology. Applied Energy. 2018;**211**: 997-1008

[11] Chen Q, Chua KJ. A spray assisted low-temperature desalination technology. In: Emerging Technologies for Sustainable Desalination Handbook. Elsevier; 2018. pp. 255-284

[12] Chen Q, Li Y, Chua K. Experimental and mathematical study of the spray flash evaporation phenomena. Applied Thermal Engineering. 2018;**130**:598-610

[13] Chen Q et al. Development of a model for spray evaporation based on droplet analysis. Desalination. 2016;**399**: 69-77

[14] Chen Q, Li Y, Chua K. On the thermodynamic analysis of a novel low-grade heat driven desalination system. Energy Conversion and Management. 2016;**128**:145-159

[15] Chen Q et al. On the second law analysis of a multi-stage spray-assisted low-temperature desalination system. Energy Conversion and Management. 2017;**148**:1306-1316

[16] Chen Q et al. Energy, economic and environmental (3E) analysis and multi-objective optimization of a spray-assisted low-temperature desalination system. Energy. 2018;**151**:387-401

[17] Chen Q et al. Energy, exergy and economic analysis of a hybrid spray-assisted low-temperature desalination/thermal vapor compression system. Energy. 2019;**166**:871-885

[18] Chen Q et al. Thermodynamic optimization of a low-temperature desalination system driven by sensible heat sources. Energy. 2020;**192**:116633

Chapter 6

Aqueous Silica and Silica Polymerisation

Lucy Lunevich

Abstract

Dissolved silica (SiO_2) is supplied to the environment by chemical and biochemical weathering processes despite the fact that dissolved silica has many stable and unstable dissolved forms (silica species). The processes involve ion substitution and chelate forming reactions which remove mineral lattice cations. The concentration of dissolved silica in natural waters is determined by a buffering mechanism which is thought to require the sorption and desorption of dissolved silica by soil particles. For instance average concentration of silica in some groundwater like coal seam gas water ranges between 0.1 and 80.0 ml/L. The dissolution process of silica and silicates from rocks into water is mainly due to hydrolysis of silica-oxygen-silica bonds, resulting in the liberation of silicic acid ($Si(OH)_4$) and silicates into aqueous phase. It is difficult to define precisely the term 'aqueous silica' as there is an array of silica species possible. Temperature, pH and ionic strength have a substantial influence on the solubility of amorphous silica and forms of silica present in a solution. This phenomenon of silica chemistry can be explained by presence of various silica species, which frequently define silica solubility and physicochemical reactions. It appears that some silica species behave as organics. For seawater the composition is relatively balanced; though, this might not explain low silica precipitation in seawater desalination.

Keywords: aqueous silica, silica speciation, silica polymerisation, silica chemistry, reverse osmosis desalination, brine treatment, coagulation

1. Introduction

In this chapter, the past and recent studies on various aqueous silica species and its impact are discussed. Behaviours of aqueous silica species were studies using reverse osmosis (RO) desalination systems and ^{29}Si NMR techniques and coagulation to gain better understanding of aqueous silica species polymerisation and practical implications of these techniques [1–6].

Practical silica solubility or the solubility defined empirically is a key for prevention of silica polymerisation in RO desalination systems [1–5, 7]. Soluble or dissolved (reactive) silica contains different forms of silica or silica species (**Figure 1**); monomer, dimers, trimmers and other polymers of silicic acid in different solutions [5, 7, 8, 12]. These dissolved silica species ($Q^0 < Q^1 < Q^2 < Q^3$ aggregation) (**Figure 1**) can be presented in various ionisation states which depend on the pH of solutions and silica concentrations and presence of other anions and cations [1, 2, 5, 8, 10–16]. Chemical reactions between these silica species and cations and

IntechOpen

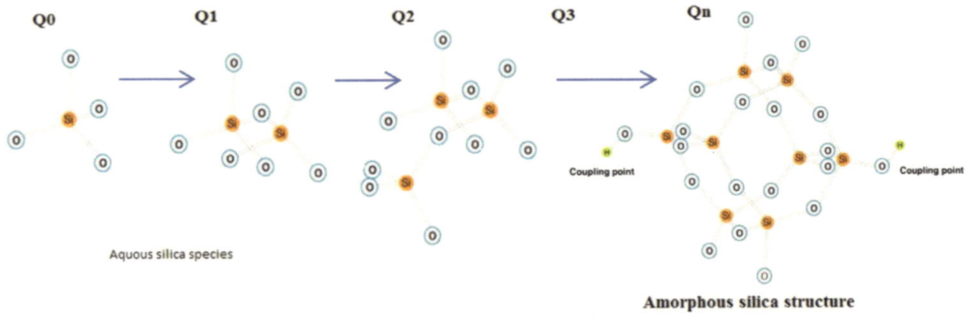

Figure 1.
Dissolved silica species ($Q^0 < Q^1 < Q^2 < Q^3$ aggregation) polymerisation path into amorphous silica structure (mechanism proposed by Iler [14]).

anions often present in the Concentration Polarisation (CP) layer in super-saturation conditions during the reverse osmosis chemical separation of water molecules are commonly lead to irreversible silica scale formation on the membrane surface [17–20]. Contrarily, not all highly super-saturated silica streams lead to scaling of membrane surfaces as it was shown in the past research [1, 5, 9].

1.1 Summary

Let us start with summary of the set of conclusions which the author found in the past research [1–25]. Because aqueous silica polymerisation depends on number of oxygen atoms attached to the particular group of aqueous silica the term 'Practical' silica solubility was introduced in that research to be able more accurately define aqueous silica chemistry and polymerisation processes [5–9]. Dissolved silica species polymerisation path or aqueous silica solubility was originally described by Kiselev [10, 11] and then adopted by Iler [14] and then later developed by Bergna [12, 13] others [26, 27] (**Figure 1**).

The solubility limit for silica in various waters vary, it is estimated, however, at approximately 120 mg/L at 25°C [5, 28–30]. Practical means an experimental silica solubility to verify the silica solubility in specific conditions – physical and chemical [1, 2, 5, 31]. The solubility defined empirically is also a key for prevention of silica polymerisation and silica scale formation in RO desalination systems [5, 32–43].

According to Sjorberg, Lunevich, and others soluble or dissolved (reactive) silica contains different silica species; monomer, dimers, trimmers and other polymers of silicic acid in different solutions [43–51]. These dissolved silica species can be presented in various ionisation states which depend on the pH of solutions, silica concentrations, presence of other silica species and other cations and anions [3, 5, 43–51]. During the reverse osmosis chemical separation of water molecules (separation of salts) dissolved silica accumulate on the membrane surface which could lead to silica scale formation on the membrane surface.

Interesting that not all highly super-saturated silica streams lead to scaling of membrane surfaces [1, 3, 5]. According to Lunevich in high salinity waters sodium could prevent silica deposition on the membrane surface [1, 5]. The phenomenon of silica chemistry can be summarised as follow:

i. Silica precipitation profiles (various silica concentrations and pH conditions) studied in deionised water showed that rapidly increasing silica concentrations in the recycled stream had little impact on flux decline likely as the result of reversible hydrolysis and condensation processes [5].

ii. The results documented in the research [1–3, 5] shown that NaCl in relatively high concentrations (8–60 g/L) depress silica solubility limits showing silica fouling at 90–81 mg/L, but no silica deposition or single silica bonding to the membrane surface was found on the membrane surface suggesting that precipitated silica remained in the recycled stream, again the indication on various silica polymerisation processes via silica species which were formed during the specific conditions, silica concentrations, pH [1–6].

iii. The maximum and stable residual silica concentration concepts were introduced to highlight potential theoretical silica solubility deviations in different water matrices and recorded pseudo-silica solubility [5] again leading to new research on silica species. Residual silica concentrations can increase above the silica solubility limit, to form super-saturated solutions, especially for low salinity and high flux conditions where silica concentrations increase rapidly in accordance with the RO concentration factor [19, 20, 52–54]. However, this result does not explain what silica species is present there. The kinetics of precipitation is slow due to the requirement for nucleation to occur and the relatively short time of the desalination experiments. It is understood that different experimental conditions could lead to formation of different silica species which are more or less resistant to precipitation and membrane scale formation.

iv. The presence of minor (7–11 mg/L) to moderate (27.7 mg/L) concentrations of aluminium in CSG waters increased silica fouling and deposition as aluminium silicate on the membrane surface dramatically [2, 3, 5]. The deposit formed on the membrane surface was much harder to dissolve compare to other silica deposit found in the experiments [5].

v. Silica precipitation occurred at pH 3 condition in both synthetic and natural CSG waters—the phenomenon which requires to further study [2, 3, 5]. The practical implications of the results of silica study obtained in the past study [5, 52] can be summarised as follows:

vi. Silica present in medium concentrations (20–45 mg/L) in CSG waters increased scale formation on the membrane surface, and therefore residual aluminium concentrations need to be minimised in the RO pre-treatment stages [5, 52]. This remains as a significant problem for industries [5]

1.2 Key findings

The research was undertaking in three prolonged phases to study potential impacts of various silica species and their forms on silica solubility and silica scale formation [5, 23, 52–57]. Comparison of the results obtained by different techniques (reverse osmosis, ^{29}Si NMR spectroscopy and coagulation) on the effect of sodium, aluminium and pH on silica polymerisation and silica precipitation patterns the following conclusions can be drawn as follows, specifically from the ^{29}Si NMR study described in the study [5]:

a. As a result of dilution with deionised water, dissolved silicate species gain (condensation process) and/or lose (hydrolysis process) monomeric silicic acid (Q^0 type surroundings). A gradual decrease of ^{29}Si NMR spectrum intensity, increased proportions of Q^0 and decreased Q^1, Q^2 and Q^3, indicates that hydrolysis or dissolution of monomeric silicic acid occurred immediately.

Overall, a consistent proportional decrease of ^{29}Si NMR spectrum for Q^1, Q^2, and Q^3 indicated that there was an equilibrium between species at the Si/M molar ratio 1.7 [2, 3, 5].

b. Here it is shown that addition of sodium chloride slightly increased the release of monomeric silicic acid (Q^0) and decreased hydrolysis of more complex silicate species (Q^2, Q^3) compared to similar dilutions with deionised waters. It suggests that sodium prevents an access of water molecule into more complex silica species structures like Q^2, Q^3 [3, 5, 56–60].

c. The effect of sodium ions on silicate species indicates that Q^0 type silica species Si-OH is the preferred attraction bond over Si-O-Na, while for Q^1 and Q^2 type surroundings Si-O-Na is preferred over Si-OH. For Q^3 type both reactions with Na^+ and OH^- may occur in combination and mix of bonds depending on the pH, and presence of stable and non-stable more complex silicate species [3–5, 31, 61]. The effect of sodium ions on silica species indicates a slowdown of the hydrolysis process.

d. There was a clear indication of a structural shift of silicate species towards condensation (precipitation reaction) processes in presence of aluminium in concentrations higher then >7 mg/L [3–5]. The presence of aluminium on silicate species has the following effects: (a) aluminium ions connected to silicate, Al-O-Si-O, resulting in a loss of sensitivity of ^{29}Si NMR spectrum, (b) aluminium forced re-arrangement of species perhaps into polymerisation and precipitation, which are also lead to a loss sensitivity of ^{29}Si NMR spectrum, and (c) it is likely some silicate species Q^1, Q^2 type precipitated as aluminium silicate.

e. A significant impact of minor concentrations of aluminium (>7 mg/L) into relatively rich in silicon sodium silicate solutions is an indication that low aluminium concentrations have a significant impact on dissolved silicate species which assume to be present in the solution in non-stable forms [2–5, 52].

f. The chemical shifts recorded at low pH 2 and 3 illustrate the presence of monomeric silicic acid, have not been found in other studies [23, 35, 66] probably due to rapid particle formation at these pH conditions and due to low interest in this experimental data from a practical application perspective [3, 5, 52].

Looking into silica behaviours as simultaneous charge neutralisation and sweep coagulation of silica is a more complex process than just incorporating the mechanisms of collision and particle growth [5]. The results demonstrated that the concentration of sodium chloride in solution inhibits the removal of silica by coagulation [2, 3, 5, 8]. In this study [5] the new hypothesis was introduced reflecting that aluminium (Al^{13+}) can substitute for sodium ions to neutralise and bridge silica sol [2, 3, 5, 12]. The relationships between $Si(OH)_4$ and Al^{3+} and Na^+ seems to play a key role in effective silica removal by coagulation [5], the details described by Lunevich [2, 3, 5]. In summary, these fundamental knowledge suggest the practical implication for the operation of a full-scale coagulation pre-treatment process in a RO plant for various groundwater as follows [5]:

1. Reduction of dissolved silica by coagulation can be relatively effective treatment via precipitation and precipitation mechanisms of dissolved silica species if an effective coagulant is selected and coagulation process is optimised [2, 3, 5].

2. It appears that pre-hydrolysed coagulant as ACH (PACI) is an effective coagulant in silica removal; however, residual aluminium concentrations could increase dramatically and required careful monitoring of the downstream concentrations and as mentioned the above optimisation of coagulation processes [3, 5, 61–70].

2. Silica chemistry and its effect on RO chemical separation

2.1 Aqueous silica (dissolved silica)

A major observation was made in the past by El-Manharawy [17–20], Gabelish [32, 33], about silica scale deposition in RO desalination for different water matrix Gabelish [17–20], El-Manharawy [32, 33]. In desalination of groundwater and coal seam waters [1–5], however, silica tends to precipitate on the membrane surface, but in desalination of seawater silica precipitates to a lesser extent El-Manharawy [18, 19]. Why? This phenomenon of silica chemistry can be explained by presence of various silica species, which we call aqueous silica or dissolved silica species we discussed the above. The silica chemistry for those waters as all groundwater is difficult to predict as composition could rapidly change [2, 3, 19, 20]. Those various silica species (**Figure 1**) are frequently define silica solubility and physicochemical reactions as it was verified by Kiselev and others [1–5, 7, 12]. Because silica in seawater consumed by various biological species or accumulated in seawater as the results of biological life the composition of silica species is relatively balanced and frequently in amorphous state (**Figure 2**); though, this might not explain low silica precipitation in seawater desalination [17, 18, 21, 22].

Dissolved silica is supplied to the environment by chemical and biochemical weathering processes which involve ion substitution, formation of various silica species and chelate forming reactions which remove mineral lattice cations. The concentration of dissolved silica in natural waters is controlled by a buffering mechanism (well known in biochemical science) which is thought to involve the sorption and desorption of dissolved silica by soil particles and sediment [22–26].

The dissolution process of silica and silicates from rocks into water is mainly due to hydrolysis of silica-oxygen-silica bonds, resulting in the liberation of silicic acid ($Si(OH)_4$) and silicates into aqueous phase [27–32]. Many suggested that It

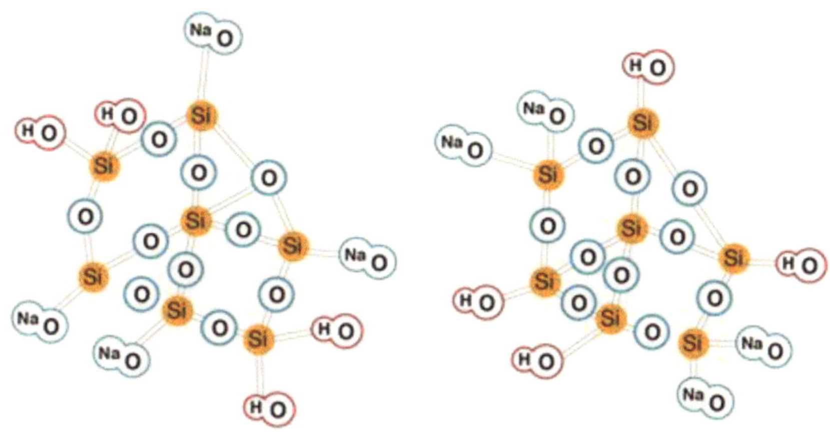

Figure 2.
Structure of two typical $Si_7O_{18}H_4Na_4$ molecules present in in the concentrated sodium silicate which provides baseline for amorphous silica [1–3, 5].

is difficult to define precisely the term 'aqueous silica' as there is an array of silica species possible [8, 12, 13, 20, 21, 32, 33].

Temperature, pH and ionic strength have a substantial influence on the solubility of amorphous silica and forms of silica present in a solution [2, 8–13]. Silica may occur in natural waters in different forms linked to special terminology as follows:

- 'Soluble' or 'dissolved' silica [8, 12] containing monomers, dimers and polymers of silicic acid, again some silica species are always in transition depending on pH and concentration of other silica species [5].

- 'Insoluble' or 'colloidal' silica [8, 12, 57], which results from polymerisation of silicic acid forming particles and three dimensions gel networks (not discussed in great details in this chapter).

- 'Reactive' silica containing monomers and dimers forms that react with ammonium molybdite within 10 min and other forms are called as 'non-reactive' silica [21, 26].

Silica polymerisation is simple a structural growth process (**Figures 1** and **2**) [1–3, 8, 10, 12]. This process leads to the formation of 'colloidal silica', which is a complex and amorphous product [8, 12, 21]. When silicate ions polymerise (condensation process), they form rings of various sizes, cross-linked polymeric chains of different molecular weights, and oligomeric structures. The arrangements of $[Si(O_4)]^{4-}$ and $[SiO_6]^{8-}$ and the tendency of these units to form a three-dimensional framework structure are fundamental to silica crystal chemistry and are studied by Smolin, Kiselev in greater details [11, 12, 23, 25, 28].

Aqueous silica sols and silica species are of particular interest in colloidal silica science because their coagulation-dispersion behaviour is said to be 'anomalous', that is, their stability in terms of electrolyte—pH control does not follow the pattern followed by almost all other oxide and colloidal materials [37–40]. To date, there has been little agreement on what constitutes stability for aqueous silica [2, 3, 8, 12]. One of the unexpected properties of silica is that silica, unlike other oxides, will not regulate charge during the approach of two surfaces [23, 37]. An explanation for the 'anomalous' behaviour of silica sols as defined by Healy [37] can be related to steric stabilisation effects that require oligomeric or polymeric silica species be present at the silica-water surface and that steric repulsion results during overlap of such layers [8, 23, 37]. At high SiO_2/Na_2O ratios, polymerisation leads to the formation of polysilicate species containing silica polymerised structures which includes 6–8 silicon atoms and consisting of predominately dissolved silica groups (Q^0, Q^1, Q^2, Q^3, Q^4 type surroundings). Very limited number of research on silica species have been conducted so far was to identify as many silica species as possible (Kiselev, Lunevich, Dietzel, Sjoberg, Marsmann, Silver).

2.2 Dissolved silica species

Aggregation (polymerisation) of dissolved silica species into more complex networks under various physical and chemical conditions leading to silica precipitation might be considered a key to understanding of silica scale deposition on the RO membrane surface when groundwater or coalseam gas water used as the feed water [1–5]. Past results from others on silica structural evolution obtained by ^{29}Si NMR show that structural control of silica polymerisation processes is complicated because many and diverse variables affect concurrent reactions differently [41–46]. Inductive and steric factors contribute to the reaction rates [34, 42, 43]. pH is probably the single most important variable in these reactions. Lunevich [2, 3, 5] and Markides [46, 47] demonstrated by ^{29}Si NMR that for all pH values from 2.5 to 11.5. The smallest particles were of a similar size being only a few nanometers in diameter [2, 3, 5], and the rate of formation of such particles gradually increased with pH (from pH 2.5 to pH 11.5). This agrees with literature models of particle nucleation and growth by Iler [12] and Bergna [50]. The effect of pH, sodium chloride concentration and presence of other cations show a different connection between the rates of aggregation, precipitation and gelation have been summarised in this chapter previously [4, 5, 51]. None of these arguments are definitive due to the nature of silica species, although assignments to monomer, dimer, etc. have been made and speculations are put forward regarding other sub-species present by a number of studies [1–40].

Dissolved and amorphous silica species can be found in commercial sodium silicate solutions, which makes it very attractive to study various silica species (**Figures 1** and **2**). Moreover, sodium silicate solutions consist of two domain states of silica—the colloidal domain (amorphous SiO_2) (**Figure 3**) illustrates two silica conditions—the dissolved silica species (mononuclear domain) and colloidal (amorphous silica—insolubility domain) silica and also illustrated in **Figure 1**. Mononuclear sodium silicate solutions contain a network of $[SiO_4]^{4-}$ species, which exist in the sodium silicate solution in equilibrium with amorphous silica (**Figure 2**).

Figure 3 illustrates that the mononuclear domain or mononuclear wall follows the line characterising $[Si(OH)_4]$ up to a pH of approximately 9, and then ionisation of dissolved silica species dramatically increases as soon as the pH 9 value is past. Within the interval between pH 9 and pH 11.5 the concentration of mononuclear silica and other silica species increase dramatically. This domain in particular is important for the study dissolved silica species (aqueous silica) by ^{29}Si NMR in particular: the monomeric domain where mononuclear Si species $[Si(OH)_4$, $Si(OH)_3$ and $Si_2(OH)_2{}^{2-}]$ prevail thermodynamically.

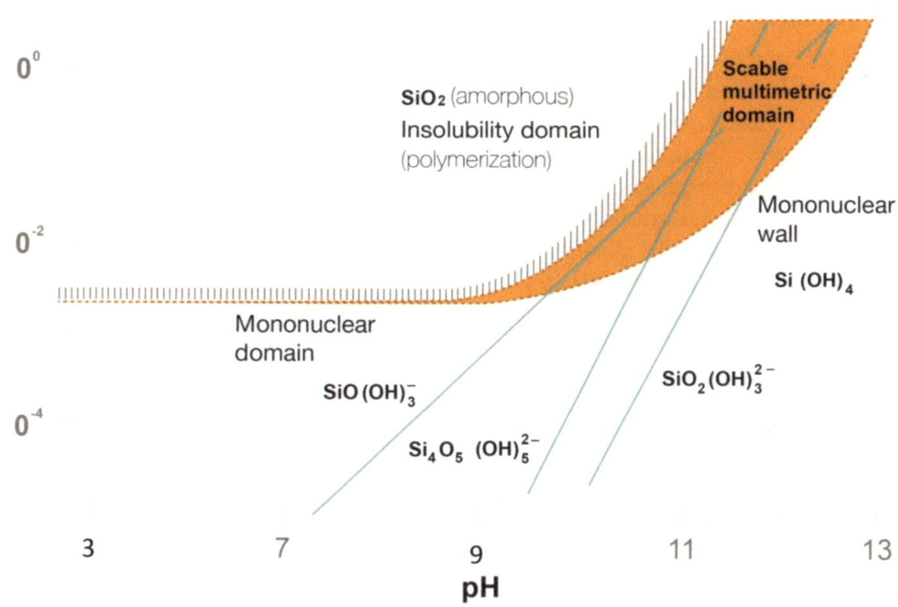

Figure 3.
The Stumm and Morgan diagram where silica species present in equilibrium with amorphous silica depending on pH and silica concentrations [5, 25].

2.3 Colloidal silica and silica species

Colloidal silica is another phenomenon of formation of various silica species during hydrolysis and condensation processes. Silica is a primary cause of concern for fouling RO membranes in desalination systems [52–56]. In the presence of carbonic acid (H_2CO_3), silica has two acid–base characters that impacts the characteristics of the silica and its interaction with membrane surfaces. First, the complexation of silica with hydrated forms of heavy elements (calcium, aluminium, magnesium and iron) creates colloids that grow through polymerisation and bridge with organic and inorganic matter to form gel-like layers on membrane surfaces the data which have been well studied by others [52–56]. Second, reactive silica is known to consist of low ionised forms (such as monomeric silica acid) at pH of 6–9 and to form an essentially all-silica gel or cake structure [52]. These silica structures cause flux decline and higher TMPs. Research has shown that colloids can be composed of any number of different materials; the mostly commonly encountered inorganic colloid is silica (SiO_2) [56–60].

Colloidal silica results from the polymerisation of silicic acid containing particles and three-dimensional gel networks [2, 12, 23, 38]. Silica may also form amorphous silica deposits especially in the presence of calcium carbonate and calcium sulphate [5, 21, 28, 56]. Colloidal silica is the result of the silico-oxygen acid polymerisation process [12]. Colloidal silica can form in bulk solution or RO feed when dissolved silica solubility exceeds the silica solubility limit.

In CSG waters, silica exists as either colloids or as un-dissociated (ortho-) silicic acid (H_4SiO_4) when the pH is between 8.5 and 9.2. A second form of silica foulant is silicates, which are complex forms of silica in which hydroxides of other elements copolymerize with silicic acid [10–12, 57]. Therefore, silica fouling may be mitigated to some extent through pre-treatment of the raw water by coagulation [5, 37]. Ideally, coagulation should leave no aluminium, no silica, and no ferric ions in the pre-treated water for RO feed [5, 37].

2.4 Silica polymerisation

Silica polymerisation and deposition, on the RO membrane surface, has been researched experimentally [54–56] and more recently computational simulations have been performed [2–8, 17–20] where the molecular mechanism and rate of hydrolysis have been explored through calculation of the reaction barriers and pathways the main factors influencing silica polymerisation are pH, temperature, saturation, impurities present in the solution, and the autocatalysis effect of already precipitated silica that could accelerate further precipitation and scale formation [5, 8, 12, 23, 56]. As can be observed total silica surface area in solution is also a factor determining the rate of silica polymerisation [11, 12, 57].

Computational simulations found that the siloxane bond, often presented on silica surfaces, is difficult to hydrolyse because of the high reactivation energy barrier, especially with the aid of hydrolysis [11, 57, 79]. However, monomers. However, monomers of silicic acid condense to form larger oligomers, which link together to produce primary particles (nucleation) [5, 79]. Depending on process conditions, these particles can either grow by reaction with monomers or grow by aggregation [8, 9, 12]. Aggregation can lead to gelation of the colloidal suspension, but not necessarily to silica deposition on the membrane surface [5, 8, 9, 37].

2.5 Silica species: scaling phenomena

Flexibility in silica solubility limit makes it difficult to control the scale formation as exact limit is unknown due to many variables and in particular silica species present. Membrane scaling phenomena are governed by the silica solubility limit prevailing in the CP layer on the membrane surface [54–56]. Semiat [54–56] suggest that the rate of change in the silica scale formation during the course of RO processing is dictated by two opposing trends: the concentration effect due to permeate withdrawal and increasing osmotic effects which acts as to decrease the rate of silica scale deposition. Permeability decline data provides a more accurate characterisation of the silica scaling process [5, 54–56]. What is not yet confirmed in the experimental silica studies by Semiat [54–56] is that the impact of dissolution (hydrolysis) on existing silica deposits. This study was in the detail discussion by others [2, 3, 5]. Will silica deposit on the membrane surface as a result of monomer silica groups form colloidal silica structure and concentrations exceeding the practical solubility limit or will it remain in dispersion because hydrolysis and condensation processes reversible.

Baoxia [61] and Elimelech [53] studied silica scaling reversibility in RO process by proposing three steps of both homogeneous and heterogeneous nucleation processes on the membrane surface. The mechanisms proposed by them of silica precipitation leading to two different nucleation processes are conflicting to what can be expected for homogeneous silica nucleation [5, 8, 57] and discussed by Bergna and Ilier [10, 12]. The diagram (**Figure 3**) describes indicative silica species distribution in different silica solubility zones, for various pH and concentrations. These silica species ($Q^0 < Q^1 < Q^2 < Q^3$) arise from published light-scattering experiments and was well documented by Kiselev and others [8, 9].

Figure 4 illustrates a typical silica polymerisation path proposed by Iler—increasing dissolved silica concentration well above solubility limit in the CP layer will lead to silica precipitation likely firstly by aggregation of monomeric silica acid in bulk solution (colloidal state) before deposition on the membrane surface. Dietzel [21–25] also illustrated in his analytical study of dissolved silica polymerisation pathways that dissolved silica species in the polymerisation process follows as $Q^0 < Q^1 < Q^2 < Q^3$ aggregation path until they react with other impurities in the other words the silica polymerisation path distracted [2, 5]. It is possible, however, the presence of other potential coupling points on the membrane surface such as –OH, –COOH groups and coupling points that are a function of the water chemistry from precipitates such as $Al(OH)_n$, $Fe(OH)_n$ leading to monomeric silicic acid coupling to the membrane surface [5, 37]. Therefore, the chemical processes of water desalination by RO (Dehydration Model) are discussed in greater details by El-Manharawy and Hafez [15–18].

2.6 Kinetics of silica polymerisation

According Lunevich, Bergna and others a considerable amount of study has been devoted to the polymerisation of silicic acid, but little work has been done

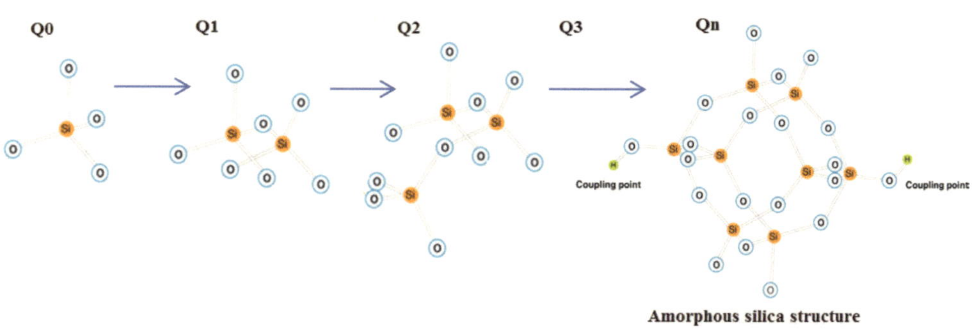

Figure 4.
Dissolved silica species polymerisation ($Q^0 < Q^1 < Q^2 < Q^3$ aggregation) path (mechanism proposed by Iler and confirm by others [5, 8, 9, 12, 57].

on the understanding of kinetics involved in the process of polymerisation of silica on RO membrane surface and colloidal silica layer formation [2, 3, 5]. Semit [54–56] and Sheikholeslami [59] have studied RO feed solutions and precipitation of silica in concentrations approaching those found in synthetic and natural waters. In an examination of an aqueous solution of silicic acid in the pH range 7–10, the rate of disappearance of monomeric silicic acid was found to follow third order kinetics. Kiselev [8] and Zhuravlev [78] offer similar results. The third order kinetic behaviour of silica polymerisation has also been noted by Marshall [70–76], in acid solutions, but as the pH of the system is increased there was a noticeable change in mechanism. It appears, however, the mechanism of polymerisation of silica species or silica systems of low concentrations is not completely understood [15–17, 19, 54–56]. It seems in different silica concentration ranges, the silica polymerisation rate is quite different [5]. Again Kiselev [8] and Zhuravlev [78] offer similar results. For instance, the polymerisation reaction of monomeric silicic acid in the presence of base was found to follow second order kinetics. Iler [12] and Kiselev [8, 9] and Smolin [60] have reported that the polymerisation process follows second order kinetics in in relatively low silica concentrations 1.8–3.0%. Semiat [54–56] and Ning [58–59] studied silica in the relatively lower concentration range 0.2–0.5% and reported that the reaction was first order with respect to SiO_2 and first order with respect to hydroxide and it was enough time for silica precipitation. It seems many agreed that the two species are necessary for the polymerisation reaction to take place are a silicic acid anion and a neutral silicic acid molecule [5, 8, 9, 12, 57, 78]. As the two reactants approach each other in the solution, it is possible the first reaction involves the formation of a hydrogen-bonded intermediate which leads to formation various silica species. The hydrogen bond formed would allow the reactants to be held in close proximity, so that the splitting out of a hydroxyl with subsequent formation of a silicon-oxygen bond can occur. However, it is unknown which silica species will be formed and how many stable and unstable silica species. This mechanism is controlled, as would be expected, by the ionisation of silicic acid which in turn depends on the pH of the system [5, 37]. Bishop [45] and Greenberg [8, 66] confirmed that the polymerisation occurs through one oxygen bridge and the system appears to form only in linear chains [8, 12, 57, 79].

The kinetics of silica polymerisation in dilute aqueous solutions was also studied by Weres, Smolin [60], Marshall [28–32], and Semit [33]. They all found that the state of ionisation of the silica surface controls the rate of polymerisation. It has been confirmed that the rate of deposition of dissolved silica on the surface of amorphous silica is proportional to the surface density of ionised silanol groups [62, 65, 66, 79]. The extent of surface ionisation also determines the value of the surface tension, and this also the rate of homogeneous nucleation [79]. Applin [79], Lunevich [5] Zhuravlev [78] found that the 'polymerisation rate' increased rapidly with increasing dissolved silica concentration in solutions, and with increasing dissolved salt concentration at pH 3.

2.7 Impact on NaCl on silica species and silica polymerisation

The results recorded in the dilution with H_2O study illustrate that monomeric silicic acid may exist as isolated molecules, so called monosilicic acid Q^0 type surroundings or ($Si(OH)_4$) and Q^1 type or $Si_2O_3(OH)_4^{2-}$, (and as linked molecules, called polysilicic acid) and as Q^2, Q^3 types. Silica polymers consist of silicate tetrahedrons that are linked via silicon-oxygen-silicon bonds. Addition of H_2O to the solution immediately initiates hydrolysis of monosilicic acid groups and as a result adjustment of equilibrium between the rest of the silicate species [5].

The model of silica polymerisation in high salinity waters (10–30 g/L as NaCl) presented in **Figure 5** illustrates that the introduction sodium chloride into the mother sodium silicate solution impacts the sensitivity of ^{29}Si NMR spectrum for Q^1, Q^2, Q^3 types. The proportion of Q^3 type was significantly reduced immediately after the first addition of sodium chloride at Si/M molar ratio 1.41, presumably due to shielding effect of sodium ions. Further addition of sodium chloride into the studied solutions, Q^1 and Q^2 type surroundings disappeared, and Q^3 type also gradually decreased. This is consistent with the hypothesis that sodium ions tend to surround polymeric silica species (**Figure 5**) acting as a stabilising agent, protecting polymeric species from attack by water molecules [2–5].

The research with sodium dilution showed significant increased proportions of monomeric silicic acid (Q^0) compare to similar dilutions with H_2O, [2, 3, 5]. This is a result of sodium ions attracting water shells, which initiate separation of monomeric silicic acid from polymeric silica groups.

When concentrations of dissolved silicate species exceed its solubility limits the nucleation process will, in principle, be governed by interacting silanol groups that interact to form –O–Si–O–Si–O bonds [5, 8, 9, 12]. This is a fundamental reason why practical silica solubility is necessary to define for any specific process conditions as it was studied for RO systems by many discussed in this chapter. Under these conditions, the probability of interactions between neighbouring silanol groups to form –Si–O–Si– bonds is higher, and therefore intramolecular nucleation would be favoured [5, 12, 52].

When silicate precipitation the monomer groups $[SiO_4]^{4-}$ and $[SiO_6]^{4-}$ randomly pack and rapid growth results in a non-periodic structures (**Figures 1** and **2**) [5, 8, 12, 54]. In medium to high salinity waters, however, dissolved and already polymerised silicate species will be surrounded by sodium ions as shown in **Figure 5**. Sodium ions attract to silicate and in particular silica monomers $[SiO_4]^{4-}$, but also can interact with –OH groups present on the membrane surface in abundance according to the Dehydration Model by El-Manharawy. It appears, however, the Na–O–Si–O–Si–O– attractions is stronger then –O–H–O–Na attractions [23, 42–46]. According to Lunevich then more silicon atoms present in silicate polymeric structure the stronger (equal) the sodium binding structure will be formed.

As can be seen from **Figures 1** and **2** polymeric silicate species will aggregate into colloidal silica in bulk solution. The polymerised silicate is unlikely to deposit

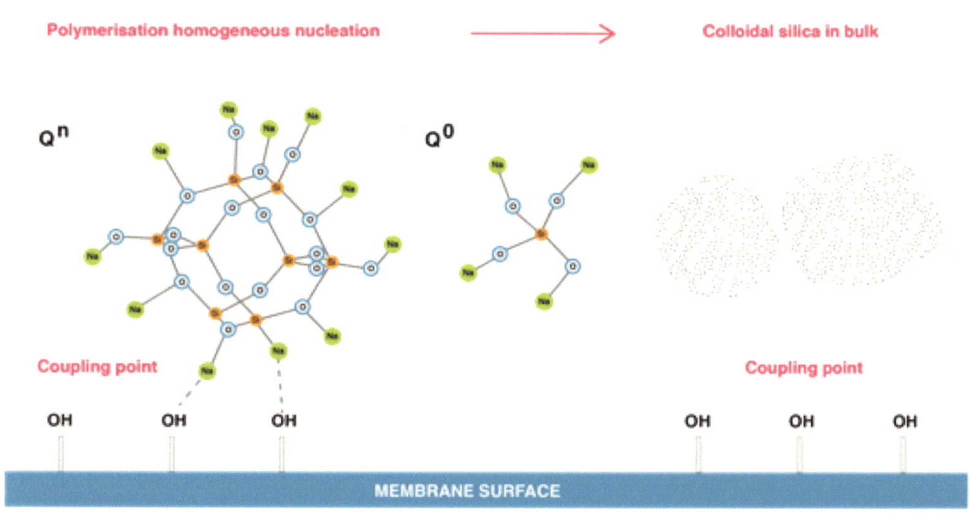

Figure 5.
Impact of sodium on silica polymerisation high salinity waters without presence of aluminium [5].

on the membrane surface due to sodium ions creating barriers between –OH groups and silicate. Monomeric silicic acid can potentially deposit solely on the membrane surface, but the reverse of this process will be apparent as it is likely dissolution (hydrolysis) process will dominate for monomeric silicic acid (**Figures 1** and **2**) [2, 3, 5]. It is know that monomeric silicic acid can coat natural organic matters presented on membranes, in this case silicate deposition on membrane surfaces is possible [72, 73, 78]. Healy [64–69] suggests that sodium ions have a number of effects on silicate species. For instance Na^+ ions in small concentrations (<8 g/L) attract water molecules and sustain further dissolution (hydrolysis) of monosilicic acid increasing the concentration of Q^0 type and at the same time preventing access to polymeric silicate structures such as Q^2, Q^3 types.

The effect of aluminium ions on silicate species, shown in **Figure 6**, is reduced ^{29}Si NMR spectra peak proportion due to silicate precipitation as aluminium silicates. It appears aluminium ions can over time disassemble polymeric silicate structures due to $(AlO_4)^{-5}$ having similar bonds to Si with oxygen O=Al—O=Al—O—, so Al can easily access polymeric silicate species leading to an irreversible process of scale formation on the membrane surface as shown in **Figure 6** [2, 3, 5].

Opposite to the effect of sodium ions on silicate species, aluminium ions force re-arrangement of silicate species leading potentially to precipitation (**Figure 6**). This is represented in **Figure 6**, when major peaks representing Q^1, Q^2, Q^3 types transformed into small, multiple peaks. It appears that in presence of aluminium ions, monomeric silicic acid and other dissolved silicate species can deposit on the membrane surface without following the classical polymerisation path ($Q^0 \rightarrow Q^1 \rightarrow Q^2 \rightarrow Q^3 \rightarrow Q^4$). Silica polymerisation does not occur in the order described by Iler and others [8, 12]. Instead it appears as silica polymerisation occurring via monomeric silica species (**Figure 6**). It is likely occurs in water when a number of conditions are present for silicate to deposit as monomeric silicate, such as elevated residual aluminium [3, 5, 77–79].

As many recognise [8, 9, 12, 35, 57] silicate has a special relationship with sodium and aluminium. Sodium tends to surround silica species preventing it from polymerisation [2, 3, 5]. Aluminium seems to be able to break silica species [2, 3, 5, 8, 12] and change silica precipitation patterns as it was found in the recent study [5].

The effect of aluminium on dissolved silica species recorded in the research [5] showed that aluminium ions can interact with dissolved silicate species such as Q^2, Q^3 types forcing re-arrangement of species (**Figure 6**). Aluminium ions can serve as coupling points for polymerised silicate to deposit on the membrane surface as shown in **Figure 6**, leading to heterogeneous silicate polymerisation [5, 8, 9, 12, 57]. Both sodium and aluminium are present in CSG waters as other groundwaters and

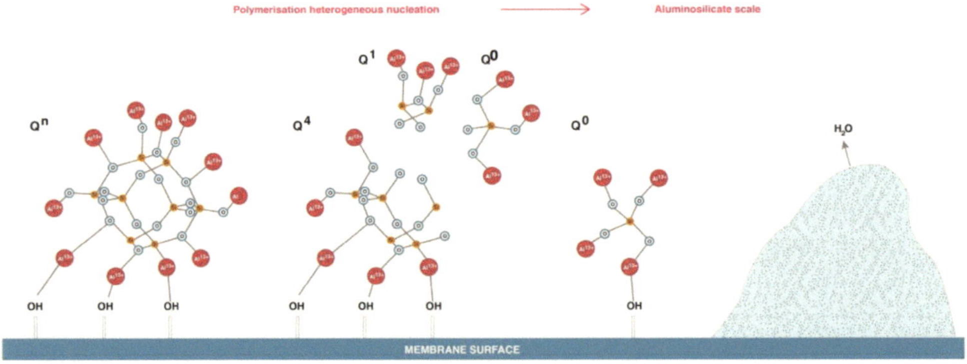

Figure 6.
Impact of aluminium ions presence on silica species polymerisation and deposition on RO surface.

affect the speciation of silicates and precipitation patterns as it can be concluded from the results recoded here [2, 3, 5]. The relative proportion of silicate species, such as Q^0, Q^1, Q^2 and Q^3 type surroundings, is strongly pH-dependent [79], and controlled by different processes such as hydrolysis (dissolution), condensation (precipitation) and complexation (aggregation) [70–72].

Structural bonding of the silicate species, or how many silicon atoms are bound to other silicon atoms via oxygen, is complex because diverse variables affect concurrent reactions in different ways [5, 8, 12, 57]. The silica system such as –O–Si–O–Si–O– linkages, however, is often considered as the key reaction dominating many geochemical processes [8, 12, 57]. The bonds play a key role in many silicate transformations [8, 12, 56]. The hydrolysis of this siloxane bond in the absence of defects has been studied by Walsh and Wilson [12], Marshal [70–76] Cypryk [12] and Pelmenschikove [12, 65–76]. These studies indicate that the high activation energy barrier effectively makes this kind of hydrolysis unlikely at ambient conditions and requires confirmation of silica solubility in specific process conditions.

2.8 Conclusion

In this chapter, results the study various aqueous (dissolved) silicate species are discussed. For the first time a method was developed to evaluate the impact of sodium and aluminium cations on dissolved silicate species using three different research methods. The discussion highlights importance of silica species in light of understanding basic silica chemistry.

Acknowledgements

The author wishes to sincerely thank Professor Stephen Gray, Doctor Peter Sanciolo, Professor Andrew Smallridge at the Victoria University, Melbourne, and Professor Tomas Healy at the University of Melbourne. The author expresses deep gratitude to Professor Raphael Semiat at the Technion, Israel Institute of Technology, Haifa for sharing his research works and Professor Jeremy Joseph at Royal Holloway University of London and dearest colleague from URS Corporation Ltd., Melbourne.

Conflict of interest

The materials dicussed in this chapter is the comprehensive summary of the past research works. All necessary permissions were obtained from the previous publishers.

Thanks

To the creator, for providing for me with a life of fulfilment and so many wonderful opportunities to grow, discover, and learn from others and to be able to contribute body of knowledge on silica and the next century of research which will be defined by silica (SO_2).

Author details

Lucy Lunevich
School of Engineering, College of Science, Engineering and Health, RMIT University, Melbourne, Australia

*Address all correspondence to: lucy.lunevich@rmit.edu.au

References

[1] Lunevich L, Sanciolo P, Dumée L, Gray RS. Silica fouling in high salinity waters in reverse osmosis desalination (sodium-silica system). Journal of Environmental Science: Environmental Science: Water Research & Technology. 2016;**2016**(2):539-548

[2] Lunevich L, Sanciolo P, Milne N, Gray RS. Silica fouling in coal seam gas water in reverse osmosis desalination. Journal of Environmental Science: Water Research & Technology. 2016;**2016**(123):53-67

[3] Lunevich L, Sanciolo P, Smallridge A, Gray S. Silica scale formation and effect of sodium and aluminium ions—29Si NMR study. Journal of Water Resources and Technology. 2016;**120**:23-45

[4] Lunevich L, Sanciolo P, Smallridge A, Gray S. On silica the edge: Silica fouling in reverse osmosis desalination. In: 2nd International Conference in Singapore. 2015

[5] Lunevich L. Silica fouling in coal seam gas water in reverse osmosis desalination [PhD thesis]. Melbourne, Australia: Institute for Sustainability and Innovation, Victoria University; 2015

[6] Gorrepati E, Wongthahan P, Raha S, Scott Fogler H. Silica precipitation in acidic solutions: Mechanism, pH effect, and salt effect. Langmuir. 2010;**26**(13):10467-10474

[7] Sancilo P, Milne N, Taylor K, Mullet M, Gray S. Silica scale mitigation for high reverse osmosis of groundwater for a mining process. Journal of Desalination. 2014;**340**:49-58

[8] Lunevich L. Asia Pacific Membrane Conference in Brisbane 28-29 November, 2012, Presentation on Silica Chemistry and its effect on Reverse Osmosis Desalination. 2012

[9] Lunevich L. Victoria University International Conference, Melbourne 23 July 2012, Presentation on Coal Seam Gas Brine Water Management and Salt Recovery. 2012

[10] Kiselev AV, Yashin Y. Gas-absorption Chromatography, Russia. New York: Plenum Press; 1956, 1959

[11] Kiselev AV. Intermolecular Interactions in Adsorption and Chromatography. Moskow: Vysshaya Shkola; 1986

[12] Bergna HE, Roberts WO. Colloidal Silica: Fundamentals and Applications. New York: CRC Press; 2006. pp. 131-128

[13] Bergna H. The Colloid Chemistry of Silica. Washington, DC: American Chemical Society; 1994

[14] Iler RK. The Chemistry of Silica—Solubility, Polymerization, Colloid and Surface Properties, and Biochemistry. New York: Wiley-Interscience; 1976. p. 866

[15] Legrand A. The Surface Properties of Silicas. Paris: John Wiley & Sons; 1999

[16] Lerman SI, Scheerer CC. The chemical behaviour of silica. In: Ultrapure Water. ResinTech Inc. 1988

[17] El-Manharawy S, Hafez A. Technical management of RO system. Journal of Desalination. 2000;**131**:173-188

[18] El-Manharawy S, Hafez A. Study of sweater alkalization as a promising RO pre-treatment method. Journal of Desalination. 2002;**153**:109-120

[19] El-Manharawy S, Hafez A. Water types and guidelines for RO system design. Journal of Desalination. 2001;**139**:97-113

[20] El-Manharawy S, Hafez A. Dehydration model for RO-membrane

fouling: A preliminary approach. Journal of Desalination. 2001;**153**:95-107

[21] Dietzel M. Geloste polymere und monomere Kieselsauren und die Wechselwirkung mit Gibbsit und Fe–O–OH– Festphasen. Germany: Habilitations—Schrift, Universitat Gottingen; 1998. p. 93

[22] Dietzel M. Dissolution of silicates and the stability of polysilicic acid. Geochimica et Cosmochimica Acta. 2000;**64**(19):3275-3281

[23] Dietzel M, Bohme G. Adsorption und stabilitat von polymerer Kieselsaure. Chemie der Erde. 1997;**57**:189-203

[24] Dietzel M, Usdowski E. Depolymerization of soluble silicae in dilute aqueous solutions. Colloid Polymeric Science. 1995;**273**:590-597

[25] Dietzel MB. Interaction of monosilicic and polysilicic acids with mineral surfaces. In: Water-Rock Interactions. Vol. 6. London: Springer; 2002. pp. 217-223

[26] Sjöberg S. Silica in aqueous environments. Journal of Non-Crystalline Solids. 1996;**196**:51

[27] Stumm W. Chemistry of the Solid-water Interface. New York: Wiley-Inerscience; 1992. p. 428

[28] Anick DJ, Ives J. The silica hypothesis for homeopathy: physical chemistry. Journal of Homeopathy. 2007;**96**:189-195

[29] Harris T. Chemical Structure of Silicates in Solutions. New York: John Wiley; 1999. pp. 630-760

[30] Dove PM, Rimstidt JD. Chapter 8: Silica-water interactions. In: Heaney PJ, Prewitt CT, Gibbs GV, editors. Silica, Reviews in Mineralogy. Journal of Mineralogy. Vol. 29. 1994. pp. 259-308

[31] Hamrouni B, Dhahbi M. Analytical aspect of silica in saline waters—Application to desalination of brackish waters. Journal of Desalination. 2001;**136**:225-232

[32] Gabelich C, Chen W, Bradley Y, Coffey M, Suffet M. The role of dissolved aluminium in silica chemistry for membrane processes. Journal of Desalination. 2005;**180**:307-319

[33] Gabelich C, Yun T, Coffey B, Suffet I. Effect of aluminium sulphate and ferric chloride coagulant residuals on polyamide membrane performance. Journal of Desalination. 2002;**150**:15-30

[34] Ning R, Troyer T. Colloidal fouling of RO membranes following Mf/UF in the reclamation of municipal wastewater. Journal of Desalination. 2007;**208**:232-237

[35] Brant J, Kwan P. State-of-the Science Review of Membrane Fouling: Organic, Inorganic and Biological. Bureau of Reclamation, US: California Department of Water Resources; 2013

[36] Semiat R, Bramson D, Hasson D. International Membrane Science and Technology Conference, No. 7, University NSW, Sydney, Australia. 1996. pp. 271-273

[37] Semiat R, Sutskover I, Hasson D. Scaling of RO membranes from silica supersaturated solutions. Journal of Desalination. 2003;**157**:169-191

[38] Semiat R, Sutskover I, Hasson D. Technique for evaluating scaling and its inhibition in RO desalinating. Journal of Desalination. 2001;**140**:181-193

[39] Ning R. Discussion of silica speciation, fouling, control and maximum reduction. Desalination. 2002;**151**:67-73

[40] Allen L, Matijevic E. Stability of colloidal silica: II. Effect of hydrolysable

cations. Journal of Colloidal Science. 1971;**35**(1):66-76

[41] Brady PV, House WA. Chapter 4: Surface controlled dissolution and growth of minerals. In: Brady PV, editor. Physics and Chemistry of Mineral Surfaces. Boca Raton, New York, London, Tokyo: CRC Press; 1996. pp. 225-305

[42] Elimelech M, Bhattacharjee S. A novel approach for modelling concentration polarization in crossflow membrane filtration based on the equivalence of osmotic pressure model and filtration theory. Journal of Membrane Science. 1998;**145**:223-241

[43] Elimelech M. Chemical Cleaning of Organic-fouled Reverse Osmosis Membranes. Denver, CO: US Bureau of Reclamation; 2005. p. 72

[44] Sjöberg S, Hagglund Y, Nordin A, Ingri N. Equilibrium and structural studies of silicon (IV) and aluminium (III) in aqueous solutions: V. Acidity constants of silicic acid and the ionic product of water in the medium range 0.05-2.0 m Na(Cl) at 250°C. Marine Chemistry. 1983;**13**:35-44

[45] Sjöberg S, Nordin A, Ingri N. Equilibrium and structural studies of silicon (IV) and aluminium(III) in aqueous solution. II. Formation constants for the monosilicate ions $SiO\ (OH)^{-3}$ and $SiO_2(OH)_2^{2-}$. Marine Chemistry. 1981;**10**:521-532

[46] Sjöberg S, Ohman LO, Ingri N. Equilibrium and structural studies of silicon (IV) and aluminium (III) in aqueous solution. II. Polysilicate formation in alkaline aqueous solution. A combined potentiometric and ^{29}Si NMR study. Acta Chemica Scandinavica A. 1985;**39**:93-107

[47] Sjöberg S. Silica in aqueous environments. Journal of Non-Crystalline Solids. 1996;**196**:51-57

[48] Sjöberg S, Nordin A, Ingri N. Equilibrium and structural studies of silicon(IV) and aluminium(III) in aqueous solution. II. Formation constants for the monosilicate ions $SiO(OH)^{3-}$ and $SiO_2(OH)^{22-}$. A precision study at 25°C in a simplified seawater medium. Marine Chemistry. 1981;**10**(6):521-532

[49] Sjöberg S, Ingri N, Nenner A-M, Öhman L-O. Equilibrium and structural studies of silicon(IV) and aluminium(III) in aqueous solution. 12. A potentiometric and 29Si-NMR study of silicon tropolonates. Journal of Inorganic Biochemistry. 1985;**24**(4):267-277

[50] Marsmann HC, Uhlig HF. ^{29}Si NMR Database System. 2001. PC based, Available from: http:\\www.silicium-nmr.de [Accessed: January 2013]

[51] Marsmann HC. ^{29}Si NMR, reference module in chemistry, molecular sciences and chemical engineering. In: Encyclopaedia of Spectroscopy and Spectrometry. 2nd ed. Academic Press; 1999. pp. 2539-2549. ISBN-10: 0123744172, ISBN-13: 978-0123744173

[52] Masaaki M, Kazuhiko Y. Correlation between the average particle diameter of porous silicas and ^{29}Si NMR spectra. Progress in Organic Coatings. 1997;**31**(1-2):153-156

[53] Ning R, Tarquin A, Balliew J. Seawater RO treatment of RO concentrate to extreme silica concentrations. Journal of Desalination and Water treatment. 2010;**22**:286-291

[54] Baoxia M, Elimelech M. Silica scaling and scaling reversibility in forward osmosis. Journal of Desalination. 2013;**312**:75-81

[55] Baumann H. Polymerisation and depolymerisation der kieselsaure unter verschiedenen bedingungen. olloid Zeitschrift. 1959;**162**(1):28-35

[56] Cornell RM, Giovanoli R. The influence of silicate species on the morphology of goethite (a FeOOH) grown from ferrihydrite ($5Fe_2O_3$ $9H_2O$). Journal of Chemical Society, Chemical Communications. 1987;6:413-414

[57] Applin KR. The diffusion of dissolved silica in dilute aqueous solution. Geochimica et Cosmochimica Acta. 1987;**51**(9):2147-2151

[58] Bright SJ. Silica and metal removal by pre-treatment to prevent fouling of reverse osmosis membranes. Journal of Desalination. 2002;**143**:255-267

[59] Sheikholeslami R, Al-Mutaz I, Ko T, Young A. Pre-treatment and the effect of cations and anions on prevention of silica fouling. Journal of Desalination. 2001;**143**:83-95

[60] Smolin Y. Structure of Water Soluble Silicates with Complex Cations. Leningrad: Institute of Silicate Chemistry of the USSR Academia of Science; 1967

[61] Birchall D. Silicon-aluminium Interactions and Biology. Amsterdam, New York: CRC Press; 1996

[62] Bouguerra W, Ali B, Hamrouni B, Dhahbi M. Equilibrium and kinetics studies of adsorption of silica onto activated alumina. Journal of Desalination. 2007;**206**:141-146

[63] Stumm W, Morgan JJ. Aquatic Chemistry—Chemical Equilibria and Rates in Natural Waters. 3rd ed. New York: Wiley-Interscience; 1996. p. 1022

[64] Healy T. Stability of Aqueous Silica Sols. Australia: School of Chemistry, The University of Melbourne; 1994

[65] Stumm W, Morgan J. Chemical aspects of coagulation. Journal of AWWA. 1962;**54**:971

[66] Marshall C, Fairbridge W. Encyclopaedia of Geochemistry. London: Klumer Academic Publishers; 2010

[67] Marshall WL. Amorphous silica solubilities—III. Activity coefficient relations and predictions of solubility behaviour in salt solutions, 0-350°. Journal of Geochimic, Cosmochim Acta. 1980;**44**:925-931

[68] Marshall WL. Amorphous silica solubilities—I. Behavior in aqueous sodium nitrate solutions; 25-3000°C, 0-6 motal. Geochimica et Cosmochimica Acta. 1980;**44**(7):907-913

[69] Marshall WL. Amorphous silica solubilities—III. Activity coefficient relations and predictions of solubility behavior in salt solutions, 0-3500°C. Geochimica et Cosmochimica Acta. 1980;**44**(7):925-993

[70] Marshall WL, Chen-Tung A, Chen WL. Amorphous silica solubilities IV. Behavior in pure water and aqueous sodium chloride, sodium sulfate, magnesium chloride, and magnesium sulfate solutions up to 3500°C. Geochimica et Cosmochimica Acta. 1982;**46**(2):279-287

[71] Marshall WL, Chen-Tung A, Chen WL. Amorphous silica solubilities V. Predictions of solubility behavior in aqueous mixed electrolyte solutions to 3000°C. Geochimica et Cosmochimica Acta. 1982;**46**(2):289-291

[72] Marshall WL, Warakomski JM. Amorphous silica solubilities—II. Effect of aqueous salt solutions at 250°C. Geochimica et Cosmochimica Acta. 1980;**44**(7):915-917, 919-924

[73] Yates DE. The structure of the oxide/ aqueous electrolyte interface [PhD thesis]. Australia: University Melbourne; 1975

[74] Sigg L, Stumm W. The interaction of anions and weak acids with the hydrous

goethite (a-FeOOH) surface. Collids and Surfaces. 1981;**2**(1981):101-117

[75] Sigg L, Stumm W. Aquatische Chemie: eine Einfuhrung in die Chemie wassriger Losung und naturlicher Gewasser. Vol. 3. Aufl. Zurich, Verl. Der Fachvereine, Stutgart: Teubner; 1994

[76] Siever R. Silicon-abundance in natural waters. In: Wedepohl KH, editor. Handbook of Geochemistry II-2. Vol. 14-I. Berlin, Heidelberg, New York: Springer; 1972

[77] Siler JL. Reverse Osmosis Membranes, Concentration Polarization and Surface Fouline: Predictive Models and 33 Experimental Verification [PhD dissertation]. University of Kentucky; 1987

[78] Zhuravlev LT. Silica surface. Journal of Colloids Surface. 1993;**74**(1):71

[79] Zhu X, Elimelech M. Colloidal fouling of reverse osmosis membranes: Measurements and fouling mechanisms. Journal of Environmental Science and Technology. 1997;**31**:3654-3662